Human-Centered Information Fusion

For a listing of recent titles in the
Artech House Electronic Warfare Library,
turn to the back of this book.

Human-Centered Information Fusion

David L. Hall
John M. Jordan

ARTECH HOUSE
BOSTON | LONDON
artechhouse.com

Library of Congress Cataloging-in-Publication Data
A catalog record for this book is available from the U.S. Library of Congress.

British Library Cataloguing in Publication Data
A catalogue record for this book is available from the British Library.

ISBN-13 978-1-59693-343-4

Cover design by Vicki Kane

685 Canton Street
Norwood, MA 02062

10 9 8 7 6 5 4 3 2 1

Table of Contents

Preface

The tradition of data and information fusion systems for applications such as military situational awareness, condition monitoring of machinery, and process monitoring has focused primarily on the use of physical sensors such as radar, LIDAR, acoustic, and seismic sensors to monitor physical objects. In the case of military systems, fusion systems have traditionally sought to observe, characterize, and identify targets such as tanks, trucks, aircraft, weapon systems, and sensors. The input data has included observations from physical sensors with limited inputs from human observers. Thus, the effort of information and data fusion (including functions such as signal and image processing, statistical estimation, pattern recognition, and limited automated reasoning) sought to transform physics-based observations into knowledge about physical objects via creation of state vectors providing information about target location, identification, and characteristics. Extensive research has focused on this problem ranging from the development of more sophisticated sensors to mathematical techniques to transform the observed signals, images, and scalar and vector data into state vectors. With some exceptions, the focus of data fusion research has been data and observation driven-that is, development of new and improved physical sensors, new methods for processing the data, and architectures that "served the data" (i.e., that started with the data input and sought to process that data to result in a common operational picture, situation display, or database of tracks and state vectors). In this approach, the human user was viewed primarily as an interpreter of the processing results (via interactions with a situation display or databases) and as a decision maker who made tactical decisions based on the evolving situation presented via a common operational picture.

There are two major trends, however, that impact this traditional view of data fusion. First, the types of targets or entities in which we are interested are no longer primarily physical. Instead of specific vehicles, sensors, and weapon systems, and so forth, we are becoming interested in the location, identity, and interactions of individuals and groups (social networks). Addressing a military threat such as improvised explosive devices (IEDs) involves not only the identification, location, and characterization of physical explosive devices and delivery vehicles (rental cars), but also networks of people who plan, design, manufacture, and deploy these devices. Thus, there is a hierarchy of physical to nonphysical "targets" sought, from physical devices, vehicles, communications devices to human networks and hierarchy of authority, intent, belief systems, cyber-connectivity, policies, and procedures. This is a "data-rich" but "model-poor" environment. While physics-based models exist for relating the observations of physical sensors to physical targets, no such models exist for nonphysical targets and for social networks. Beyond military applications, analysis of national threats such as health hazards and cyberattacks on national infrastructure, and so forth, also involves trying to identify and characterize human networks, including physical communications and virtual relationships.

The second major trend in information fusion is the emergence of two new major sources of information that have previously been relatively neglected: human observations and Web-based information. With the advent of ubiquitous cell phones (with associated GPS, image sensors, and on-board computing), we can consider formal and informal "communities of observers" that provide information about an evolving situation. Over 3 billion cell phones are currently used throughout the world. New Web sites that allow sharing of data (e.g., YouTube and Flickr), ad hoc reporting to national news networks (e.g., Yahoo!'s You Witness News), blogs, Facebook, and MySpace provide huge sources of data. While this data collection is not currently coordinated, it provides a potential source of information that we term "soft sensing." Robert Lucky, for example, recently described the concept of Internet-based information (Lucky, R., "A Billion Amateurs," *IEEE Spectrum*, November 2007, p. 96). Lucky states, "Meanwhile, those billion amateurs are taking pictures of everything on the planet and placing the images on Flickr and other sites. There are thousands upon thousands of pictures of every known place, taken from all angles and under all lighting conditions. Researchers are now using those pictures to create three-dimensional images and panoramic vistas." This information can significantly augment data obtained from traditional sensors such as unattended ground sensors, radar, airborne vehicles, and others. Similarly, Burke et al. have described the concept of participatory sensing, in which a community of observers might be tasked to provide information for applications such as urban planning and public health (Burke, J., D. Estrin, M. Hansen, A. Parker, N. Ramanathan, S. Reddy, and M. B. Srivastava, "Participatory Sensing," *Proceedings of WSW'06 at SenSys'06*, October 31, 2006, Boulder, Colorado). We acknowledge that the use of human observers is not completely new. The intelligence community has effectively used human intelligence (HUMINT) for many years—yet the rapid advances in information technology provide the opportunity for ad hoc observations from casual observers and the potential opportunity to incorporate these observations in automated fusion systems.

There are numerous challenges in accessing and utilizing such data. Examples of research challenges include: soft sensor tasking (how to effectively solicit information from civilian observers), data and knowledge elicitation (how to solicit information about target, activity, or event location, characteristics, or identity, and so forth), representation of uncertainty and second-order uncertainty, dealing with reporting and observational biases, deliberate information corruption, rumor effects, and many others. Nevertheless, this new information space becomes a very important part of the new information-fusion concepts. The U.S. Army has embraced some of these concepts via their "Every Soldier a Sensor" program. In addition to soldiers acting as sensors, civilians may also act in this manner, providing information about local conditions, activities, events, and other context-based information to improve our understanding about a situation or threat. Indeed, increasingly the first reported observations of disasters, accidents, and adverse weather conditions are "reported" by local observers posting pictures taken via cell phone.

This book presents a new view of multisensor data fusion that seeks to address these new trends and explicitly consider the active role of a human user/analyst. We view the inputs to the data fusion process as comprising three pillars: (1) traditional sensing resources ("S-space"), (2) dynamic communities of human observers ("H-space"), and (3) resources such as archived sensor data, blogs, reports, and dynamic news reports from citizen reporters via the Internet ("I-space"). The sensors in all three of these pillars need to be characterized and calibrated. In H-space and I-space, calibration issues related to motivation and truthfulness, and so forth, must be considered in addition to the standard physical characterization and calibration issues that need to be considered in S-space. Thus, the new approach explicitly considers the role of human observers as a major source of input that augments traditional sensor systems.

In addition, we consider a new role for the analyst-in-the-loop in data fusion. In this new role, the human analyst augments the traditional automated reasoning of computer-based fusion systems by explicitly using human cognition for pattern recognition (via visual and aural processing) as well as using semantic reasoning for context-based interpretation of evolving situations. The concept is to develop computer displays, use of sonification (transformation of data into sounds), and generation of semantic metadata from signals and images to allow the human user/analyst to become cognitively engaged in the inference process. Thus, humans participate on both "ends" of the fusion process: on the input side as members of a community of observers, and on the output side as engaged analysts supporting pattern recognition and semantic-based analysis. Finally, we discuss the prospect of ad hoc distributed collaboration on analysis and problem solving (crowdsourcing of analysis) via virtual world technologies.

This book does not present solutions to these new problems. However, it does seek to raise the reader's awareness about the new opportunities and challenges in human-centered information fusion.

Acknowledgments

The authors acknowledge the good humor, support, and dedication of Ms. Tracy Ray, who performed extensive work to provide the final camera-ready copy of this book. Her attention to detail greatly assisted in completing the project. In addition, the authors recognize the helpful comments and insights of the anonymous reviewer who provided excellent suggestions for improved clarity and additional references. Finally, the authors recognize the contributions of those responsible for a number of invited chapters. Guest authors include: Mr. Nathan McNeese (The Pennsylvania State University), Dr. James Llinas (State University of New York at Buffalo), Dr. Mark Ballora (The Pennsylvania State University), Mr. Jake Graham (The Pennsylvania State University), Dr. Mark S. Pfaff, Dr. Christine M. Newlon, Dr. Himalaya Patel, and Dr. Karl F. MacDorman (School of Informatics, Indiana University), and Dr. Bart Pursel (The Pennsylvania State University).

Chapter 1

Introduction: The Changing Role of Humans in Information Fusion

Extensive research has been conducted on multisensor data and information fusion primarily for military applications such as target tracking, situational awareness, and threat assessment. Traditional data fusion systems have focused primarily on fusing data from physical sensors to address physical targets. A new focus is emerging, namely, how to observe and understand the human terrain (understanding resident populations, trends, groups, individuals, and their interrelationships, as well as the traditional physical terrain). In addition, new concepts of human-centered fusion are emerging, including the use of human observers (humans as "soft" sensors), engagement of human cognition to support automated computer processing, and multiple ad hoc analysts to address complex problems. This chapter provides a summary of traditional data fusion and introduces the changing role of information fusion systems and the emergence of the human user/analyst/observer.

1.1 INTRODUCTION

The tradition of data and information fusion systems for applications such as military situational awareness, condition monitoring of machinery, and process monitoring has focused primarily on the use of physical sensors such as radar, Light Detection and Ranging (LIDAR), and acoustic and seismic sensors to monitor physical objects. In the case of military systems, fusion systems have traditionally sought to observe, characterize, and identify targets such as tanks, trucks, aircraft, weapon systems, and sensors. The input data has included observations from physical sensors with limited inputs from human observers. Information and data fusion (including functions such as signal and image processing, statistical estimation, pattern recognition, and limited automated reasoning) sought to transform physics-based observations into knowledge about physical objects via the creation of state vectors providing information about target location, characteristics, and identification. Extensive research has focused on this problem ranging from the development of ever more sophisticated sensors to mathematical techniques to transform the observed signals, images, and scalar and vector data into state vectors [1, 2]. With some exceptions, the focus of data fusion research has been data and observation driven—that is, development of new and improved physical sensors, new methods for processing the data, and architectures that "served the data." In effect, fusion system designers started with the input data and sought to process that data to result in a common operational picture, situation display, or database of tracks and state vectors. In this approach, the human user was viewed primarily as an interpreter of the processing results (via interactions with a situation display or databases) and as a decision-maker who made tactical decisions based on the evolving situation presented via a common operational picture.

There are two major trends that impact this traditional view of data fusion. The first trend is that the types of targets or entities in which we are interested are no longer primarily physical. Instead of specific vehicles, sensors, and weapon systems, we are becoming interested in the location, identity, and interactions of individuals and groups (social networks). Addressing a military threat such as improvised explosive devices (IEDs) involves not only the identification, location, and characterization of physical explosive devices and delivery vehicles, but also networks of people who plan, design, manufacture, and deploy these devices. There is a hierarchy of physical to nonphysical "targets" sought, ranging from physical devices, vehicles, and communications devices to human networks and a hierarchy of authority, intent, belief systems, cyberconnectivity, policies, and procedures. This represents a transition from observing and characterizing the physical landscape to observing and characterizing the human landscape. This is a "data-rich" but "model-poor" environment. While physics-based models exist for relating the observations of physical sensors to physical targets, limited models exist for nonphysical targets and for social networks. Beyond military applications, analysis of national threats such as health hazards and cyberattacks on national infrastructure also involve trying to identify and characterize human networks, including physical communications and virtual relationships.

The second major trend in information fusion is the emergence of two new major sources of information that have previously been relatively neglected: human observations and Web-based information. With the advent of ubiquitous cell phones (with associated Global Positioning Satellite (GPS) sensors, image sensors, and on-board computing), we can consider formal and informal "communities of observers" that provide information about an evolving situation.

Nearly 4 billion cell phones are currently used throughout the world. New Web sites that allow sharing of data (e.g., YouTube and Flickr) and ad hoc reporting to national news networks (e.g., Yahoo's You Witness News [http://news.yahoo.com/you-witness-news]), blogs, Facebook, and MySpace provide huge sources of data. While this data collection is not currently coordinated, it provides a potential source of information that we term "soft sensing." Robert Lucky, for example, described the concept of Internet-based information [3]. Lucky states, "Meanwhile, those billion amateurs are taking pictures of everything on the planet and placing the images on Flickr and other sites. There are thousands upon thousands of pictures of every known place, taken from all angles and under all lighting conditions. Researchers are now using those pictures to create three-dimensional images and panoramic vistas." This information can significantly augment data obtained from traditional sensors such as unattended ground sensors, radar, and airborne vehicles. Similarly, Burke et al. [4] described the concept of participatory sensing, in which a community of observers might be tasked to provide information for applications such as urban planning and public health. It should be noted that there is an extensive history of the use of human observations or reports with physical sensors. The intelligence community has effectively used HUMINT (human intelligence) to augment other types of sources. However, emerging trends in information technology make such human reporting commonplace and have introduced concepts of ad hoc reporting. The automated use of such an emerging source of information is one of the themes of this book.

There are numerous challenges in accessing and utilizing such data [5]. Examples include soft sensor tasking (how to effectively solicit information from civilian observers), data and knowledge elicitation (how to solicit information about target, activity or event location, characteristics, identity), representation of uncertainty and second-order uncertainty, dealing with reporting and observational biases, deliberate information corruption, rumor effects, and many others. Nevertheless, this new information space becomes a very important part of the new information fusion concepts. The U.S. Army has embraced some of these concepts via their "Every Soldier a Sensor" program (http://www.ausa.org/pdfdocs/IP_Sensor08_04.pdf). In addition to soldiers acting as sensors, civilians may also act in this manner, providing information about local conditions, activities, events, and other context-based information to improve our understanding about a situation or threat. Increasingly the first reported observations of disasters, accidents, and adverse weather conditions are reported by local observers posting pictures taken via cell phone.

This book presents a new view of multisensor data fusion that seeks to address these new trends and explicitly considers the active role of a human user/analyst. We view the inputs to the data fusion process as comprising three pillars: (1) traditional sensing resources ("S-space"), (2) dynamic communities of human observers ("H-space"), and (3) resources such as archived sensor data, blogs, reports, and dynamic news reports from citizen reporters via the Internet ("I-space"). The sensors in all three of these pillars need to be characterized and calibrated. In H-space and I-space, calibration issues related to motivation and truthfulness must be considered in addition to the standard physical characterization and calibration issues that need to be considered in S-space. We explicitly consider the role of human observers as a major source of input that augments traditional sensor systems.

In addition, we consider a new role for the analyst-in-the-loop in data fusion. In this new role, the human analyst augments the traditional automated reasoning of computer-based fusion systems by explicitly using human cognition for pattern recognition (via visual and aural processing) as well as using semantic reasoning for context-based interpretation of evolving situations. The concept is to develop computer displays, use of sonification (transformation of data into sounds) [6], and generation of semantic metadata from signals and images to allow the human user/analyst to become cognitively engaged in the inference process. Thus, humans participate on both "ends" of the fusion process: on the input side as members of a community of observers, and on the output side as engaged analysts supporting pattern recognition and semantic-based analysis.

Finally, we consider the possibility of using ad hoc dynamic groups of analysts to address problems. Thus, much like the emergence of new collaboration sites such as hive-mind interactions (e.g., MetaFilter [http://www.metafilter.com/], Wikipedia, Experts Exchange [http://www.experts-exchange.com/]), we argue that the rapid emergence of technologies such as multiplay distributed game engines and virtual world technologies such as Second Life and ProtoSphere, provides the opportunity for a new kind of collaboration. Ad hoc groups could be solicited or could spontaneously address complex problems such as environmental concerns due to contaminations, public health (spread of disease), or homeland security. This is an extension of the "crowdsourcing" concepts used in business, described by Howe [7].

1.2 TRADITIONAL VIEWS OF DATA FUSION

Historically, the bulk of funding for development of data and information fusion concepts has come from military applications. A wide range of work [8] has included target tracking and automatic target recognition (ATR), identification, friend, foe, neutral (IFFN), situational awareness, threat assessment, and related applications. Indeed, this research and U.S. Department of Defense (DoD) applications were the motivation for the creation of the Joint Directors of Laboratories (JDL) data fusion process model [9, 10], upon which much literature is based. In this section, we present a brief summary of the DoD legacy for data fusion, introduce the JDL data fusion process model, and describe related models. A summary of the state of the art of data fusion completes the section.

1.2.1 The Department of Defense (DoD) Legacy

The early history of multisensor data fusion was dominated by research, programs, and applications for military (Department of Defense) types of applications [11]. As discussed in the next section, these efforts led to the creation of the Joint Directors of Laboratories (JDL) data fusion process model [9, 10], the institution of an annual National Symposium on Sensors and Data Fusion (NSSDF), and multiple programs and implemented systems. The DoD legacy for data fusion includes the following:

- The Joint Directors of Laboratories (JDL) Process Model described in Section 1.2.2 [9, 10, 12, 13].
- Related process models, including:
    - Boyd's Observe-Orient-Decide-Act (OODA) Loop model [14, 15];
    - Endsley's situational awareness model [16];
    - Bedworth and O'Brien's Omnibus model [17];
    - The Transformation of Requirements to Information Processing (TRIP) model [18].
- A *Taxonomy of Algorithms* has been developed by Hall and Linn [19].
- A data fusion *lexicon* was originally developed by TechReach Inc., and a version is available online (www.nc2if.psu.edu).
- *Engineering guidelines* have been created for various aspects of data fusion development, including:
    - Requirements analysis for data fusion systems [20];
    - Architecture selection [21];
    - Algorithm selection [21];
    - Database design concepts [22];
    - Evaluation of specific types of techniques for data association and correlation [23];
    - Metrics for test and evaluation [24].
- Evolving toolkits: A survey of commercial-off-the-shelf (COTS) software related to data fusion processes has been performed by McMullen et al. [25].
- Extensive legacy of technical papers, books, and conferences.

  - The National Symposium on Sensor and Data Fusion (NSSDF)—proceeding of this ongoing conference are available from the Military Sensing Information Analysis Center (SENSIAC): https://www.sensiac.gatech.edu/external/index.jsf.
  - The Society for Photo-Interpretive Engineers (SPIE) hosts annual conferences on data fusion; see their Web site for information on past and planned conferences: http://spie.org.
  - The International Society of Information Fusion (ISIF) (http://www.isif.org) sponsors an annual International Conference on Data and Information Fusion. Proceedings of the first 10 years of the conference are available at the ISIF Web site.
  - An endnote bibliography of 2,500 references related to data fusion is available: www.nc2if.psu.edu.
- Training materials:
  - A set of 450 PowerPoint slides and annotations on data fusion are available from [26].
  - An online graduate level course is available on the topic of data fusion at The Pennsylvania State University World Campus: http://www.worldcampus.psu.edu/AboutUs_About.html.
  - Multiple two- and three-day seminars are available from commercial training vendors.
- Test-beds: A number of test-bed environments have been developed to support the evaluation of data fusion systems and algorithms. A survey of some of these test-beds and a simulation tool environment are described by [27].
- Department of Defense-affiliated data fusion centers:
  - Center for Multisource Information Fusion (CMIF), http://www.infofusion.buffalo.edu/.
  - Center of Excellence for Battlefield Sensor Fusion, http://www.tnstate.edu/ce-bsf.
- Numerous prototypes and deployed systems: Numerous surveys have been conducted of prototype and deployed data fusion systems; a sample of these surveys includes [7, 28]. An example of categories of military-related data fusion applications is summarized in Table 1.1.

Table 1.1

Examples of Military-Related Data Fusion Applications

Fusion Application	*Description/Focus*	*Types of Sensors*	*Inferences Sought*
Target tracking and surveillance	Detection and tracking of individual objects such as aircraft, ground vehicles	• Radar • Electro-optic sensors	Location, velocity, and trajectory of individual physical objects
Automatic target recognition (ATR) or specific emitter identification (SEI)	Use of observed attributes of a target to identify the class, type, or specific identity of a target such as an aircraft	• Radar and electro-optic (for aircraft) • Acoustic (for underwater vehicles) • Radio emissions (RF) for emitters or active sensors	Identification of target class, type, or even specific identity
Battlefield surveillance	General surveillance of a battlefield environment to determine enemy locations, identity, movements	• Ground-based sensors such as networked seismic and acoustic sensors • Radars and electro-optic sensors • Human observers • Airborne radar, electro-optic and communications sensors	General information concerning an area of interest including terrain, location of enemy and friendly units, and factors that affect courses of action such as trafficability and observability
Strategic warning and defense	Detection of indications of impending strategic actions (e.g., detection and tracking of ballistic missiles and warheads)	• Space-based sensors including infrared detectors • Nuclear detection sensors • Communications sensors	Indications and warnings of impending strategic actions, detection of precursors to missile launch, and detection and tracking of ballistic missiles

1.2.2 The Joint Directors of Laboratories (JDL) Data Fusion Process Model

In the early 1990s, a number of U.S. DoD large-scale funded efforts were under way to implement data fusion systems. A prime example was the U.S. Army's All Source Analysis System (http://www.fas.org/irp/program/ process/asas.htm). At that time, the field of data fusion was not well founded, with limited common understanding of terminology, algorithms, architectures, or engineering processes. The Joint Directors of Laboratories (JDL) was an administrative group that existed to assist in the coordination of research across multiple U.S. DoD Laboratories. The JDL established a subgroup to focus on issues related to multisensor data fusion (the formal name was the Joint Directors of Laboratories, Technical Panel for Command, Control and Communications (C^3) data fusion subpanel). Led by Frank White, this subgroup initiated discussions about data fusion terminology, processes, and techniques. The subgroup created what is now called the JDL data fusion process model. The model was originally published in a briefing (Kessler et al. [9]) to the Office of Naval Intelligence and later widely presented in a variety of papers, and used as an organizing concept for books [1, 2], national and international conferences, requests for proposals, and in some cases government and industrial research organizations.

Since its inception in 1991, the model has received several additions and revisions. Originally, the model included only the first four levels of fusion processing: object refinement (level 1), situation refinement (level 2), threat assessment (level 3), and process refinement (level 4). In 1999, Steinberg, Bowman, and White [10] published the first extension of the JDL model, adding a precursor level of fusion. Level 0 fusion involves sensor associated data and estimation. The idea of level 0 processing was to recognize the increasing role of smart sensors and associated processing at the sensor/source level. Subsequently, in 2000, M. J. Hall, S. A. Hall, and T. Tate [12] and independently in 2002 Blasch and Plano [13] extended the JDL model to include human-computer interaction-related issues that are used to control data fusion related processes (a level 5 process). However, as the dotted line around level 5 in Figure 1.1 denotes, there is still some debate in the data fusion community whether level 5 is relevant to the goal of developing a functional model of data fusion. More recently, other extensions to the data fusion model have been discussed by Llinas, who presents the case for further consideration of current data fusion issues, including distributed data fusion systems and ontology-based systems.

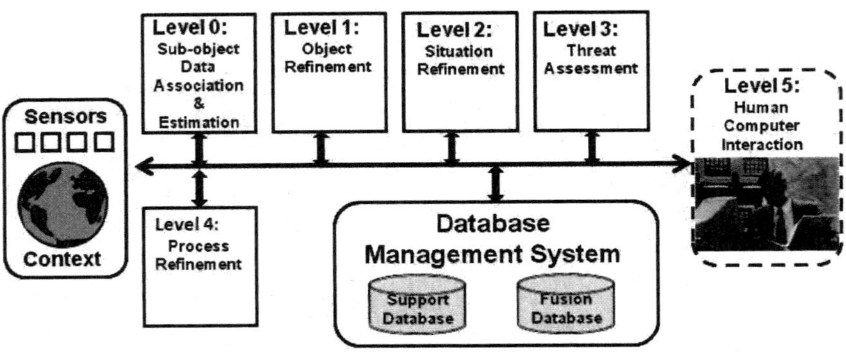

Figure 1.1 The Joint Directors of Laboratories (JDL) data fusion process model (Hall and McMullen [1]).

The six high-level processes defined in the JDL model are summarized here:

1. *Level 0 fusion (data or source preprocessing)* involves processing data from sensors (e.g., signals, images, hyperspectral images, vector quantities, or scalar data) to prepare the data for subsequent fusion. Examples of data preprocessing include image processing, signal processing, "conditioning" of the data, coordinate transformations (to relate the data from the origin or platform on which the sensor is located to a centralized set of coordinates), filtering, alignment of the data in time or space, and other transformations.
2. *Level 1 fusion (object refinement)* seeks to combine data from multiple sensors or sources to obtain the most reliable estimate of the object's location, characteristics, and identity. We speak here of an object (for example, observing a physical object such as an airplane), but we could also fuse data to determine the location and identity of activities, events, or other geographically constrained entities of interest. When fusing data related to monitoring the health of a mechanical system such as an automobile, for example, level 1 fusion would be aimed at locating and identifying possible failure or fault conditions (e.g., the location of a worn gear tooth that is disrupting a drive train). Often the issues of object/entity location (estimation) are discussed separately from the problem of object/entity identification. In real fusion systems, these subprocesses are usually performed in an integrated fashion.
3. *Level 2 fusion (situation refinement)* processing uses the results of level 1 processing and seeks to develop a contextual interpretation of their meaning. This often entails understanding how entities are related to their environment, the relationship among different entities, how they interrelate, and so on. An example of level 2 fusion is observing a mechanical system and trying to understand the current condition of the machine. Observations of vibration, noise, heat, and smell may provide an indication of whether or not a machine is functioning "normally" (and is healthy) or has a potential problem. This interpretation involves reasoning of various types, and the conclusions reached are dependent upon context. For example, an automobile that is vibrating due to high speed on a rough highway might be perceived as operating normally, while the same vehicle, making the same sounds and vibration might be considered to be abnormal if those symptoms occurred while it is operating at a slow speed on a smooth city street. The types of techniques used for level 2 fusion may involve artificial intelligence, automated reasoning, complex pattern recognition, rule-based reasoning, and many other methods.
4. *Level 3 fusion (threat refinement/impact assessment)* concerns the projection of the current situation into the future to determine the potential

impact of threats associated with the current situation. In monitoring the health of a machine, we might seek to determine how long the machine will operate safely under anticipated operating conditions. Our perception is very dependent upon context. If we are on a long trip in the middle of a lonely stretch of road, a warning light on the automobile's engine (e.g., the check engine light) might cause us more concern than if we are at home and only occasionally drive the car on short trips. Level 3 processing seeks to draw inferences about possible threats, courses of action (in response to those perceived threats), and how the situation changes based on our changing perceptions. Techniques for level 3 fusion are similar to those used in level 2 processing but also include simulation, prediction, and modeling.

5. *Level 4 fusion (process refinement/resource management)* is a metaprocess (namely, a process that addresses a process). In particular, level 4 processing "observes" the ongoing data fusion process (the other levels of processing) and seeks to make the fusion process better (more accurate, more timely, more specific) by redirecting the sensors or information sources, changing the control parameters on the other fusion algorithms, or selecting which algorithm or technique is most appropriate to the current situation and available data. The level 4 process involves functions such as sensor modeling, modeling of network communications, computation of measures of performance, and optimization of resource utilization. Level 4 processing is an optimization process.

6. *Level 5 processing (human computer interaction/cognitive refinement)* seeks to optimize how the data fusion system interacts with one or more human users. The level 5 process seeks to understand the needs of the human user and respond to those needs by appropriately focusing the fusion system attention on things that are important to the user. Types of functions may include use of advanced displays, search engines, advisory tools, cognitive aids, collaboration tools, and other techniques. This may involve use of traditional human-computer interface (HCI) functions such as geographical displays, displays of data and overlays, processing input commands, use of non-visual interfaces such as sound or haptic (touch) interfaces, and others.

A summary of the levels of fusion in the JDL model is provided in Table 1.2. Examples of types of algorithms and methods are provided in the third column. Extensive descriptions of methods for performing data fusion are provided by [1, 2, 11].

Table 1.2

Summary of JDL Model Levels

JDL Level	*Description/Purpose*	*Types of Algorithms/Techniques*
0	Source refinement—preprocessing of individual sensor or source data to improve subsequent processing	• Signal and image processing • Feature extraction • Signal and image conditioning • Predetection fusion methods
1	Object refinement—combining data from multiple sensors or sources to obtain the most reliable estimate of the object's location, characteristics, and identity	• Location estimation and tracking techniques—Kalman filters, particle filters, multiple hypothesis methods, probabilistic data association, joint probabilistic data association, random set methods • Attribute estimation and identity processing—physical models, neural nets, cluster algorithms, machine learning methods, logical and parametric templates
2	Situation refinement—level 2 processing uses the results of level 1 processing and seeks to develop a contextual interpretation of their meaning	• Pattern recognition methods—neural nets, parametric templates, cluster algorithms • Automated reasoning (e.g., logical templates, rule-based systems, Bayesian belief nets, Petri nets, intelligent agents)
3	Threat refinement/impact assessment—level 3 processing concerns the projection of the current situation into the future to determine the potential impact of threats associated with the current situation	• Same as level 2 • Course of action analysis—rapid engagement models, gaming techniques
4	Process refinement—level 4 processing "observes" the ongoing data fusion process (the other levels of processing) and seeks to make the fusion process better (more accurate, more timely, more specific) by redirecting the sensors or information sources and changing the control parameters on the other fusion algorithms	• Multicriteria optimization methods • Market-based methods (e.g., dynamic auction methods)
5	Cognitive refinement—the level 5 process seeks to optimize how the data fusion system interacts with one or more human users	• Cognitive aids—expert advisory systems, intelligent agents, knowledge-based systems, decision support tools

1.2.3 Related Models of Data Fusion

There are a number of related models that address cognitive and information processes that are related to data fusion. A survey and an assessment of these process models were conducted by [29]. A summary of the models is presented in Table 1.3 along with references that describe the models in more detail. Hall et al.

Table 1.3

Summary of Reviewed Process Models

Model	*Description*	*Field*	*References*
JDL Process Model	A functional model for describing the data fusion process	Data Fusion	Kessler et al. [9] Liggins, Hall, and Llinas [2] Hall and McMullen [1] Steinberg, Bowman, and White [10] Hall, Hall, and Tate [12] Blasch and Plano [13]
Functional Levels of Fusion	An abstraction of input-output functions of the data fusion process—focus on types of data processed and associated techniques appropriate to the data types		Dasarathy [30]
Transformation of Requirements to Information Processing (TRIP) Model	Application of the waterfall development process to data fusion—emphasis on linking inferences to required information and data collection		Kessler and Fabien [18]
Omnibus Model	Adaptation of Boyd's OODA loop for data fusion		Bedworth and O'Brien [17]
Endsley's Model of Situational Awareness	A cognitive model for situational awareness		Endsley [16]
Recognition Primed Decision Making (RPD)	A naturalistic theory of decision-making focused on recognition of perceptual cues and action	Decision-Making	Klein [31] Klein and Zsambok [32] Kaempf et al. [35]
Observe, Orient, Decide, Act (OODA) Loop	A process model of military decision making based on observing effective commanders; extended by several authors for general situation assessment and decision-making		Boyd [14, 15] Brehmer [33] Bryant [34] Rousseau and Breton [36] Grant [37]

[29] divided the models into two broad categories: data fusion models and decision-making models. To a certain extent, this is an arbitrary partitioning but reflects how these models are referenced in the literature. In addition, models such as the OODA loop have several extensions and variations. Each of these models has specific advantages and disadvantages related to describing the fusion and decision-making process. They are summarized here to indicate the potential variations in how to describe or characterize the process of fusing information to understand an evolving situation and ultimately result in a decision or action.

1.2.4 Assessment of Fusion Technology

While the intent of this book is not to provide a complete survey and assessment of data fusion, we present here a brief summary of the state of the art of data fusion, using the JDL model processes. Table 1.4 presents a summary of each JDL level process with comments on current practices and identification of some limitations and challenges. Detailed information about current practices in data fusion is provided by [2]. A brief summary of the state of each level of fusion is provided below and shown in Table 1.4.

Level 0, source preprocessing, involves a wide variety of techniques to process single-source, homogeneous data as a precursor to fusion with other data types. This includes classic signal and image processing as well as techniques for data conditioning, representation, source characterization, and related methods. Source preprocessing techniques tend to be sensor or source specific (e.g., processing of complex data such as synthetic aperture radar (SAR), inverse SAR, or hyperspectral image data). The advent of "smart sensors" with embedded processors allows significant amounts of processing to be performed (see Swanson [38]). Generation of metadata (data about the data) may range from dynamic source characterization based on the observing environment to the creation of semantic metadata describing images or signal data [39, 40]. Other work, called *predetection fusion,* has focused on combining raw signals or images from sensors or elementary features or segments of data provided by similar sensors [41]. The proliferation of sensors such as video sensors that persistently observe an environment will pose challenges for tagging the data for effective use. In addition, distributed, ground-based sensor nets require preprocessing for effective use.

A key challenge in source preprocessing involves data registration (how to ensure that sensor data about an entity can be accurately associated with a particular location, especially for image data). Brooks and Grewe [42] describe the data registration process and associated algorithms.

Level 1, entity refinement (classic target tracking and identification), involves data association and correlation, estimation of a state vector that typically represents location or kinematic parameters, and pattern recognition for target identification. Level 1 fusion is the most mature area within the JDL fusion

Table 1.4

Summary of the State of the Art in Data Fusion

JDL Process	*Current and Emerging Practices*	*Limitations and Challenges*
Level 0: Source Refinement	• Sensor preprocessing using standard digital signal and image processing methods • Detect before fuse algorithms • Automated semantic labeling of image data	• Absolute image registration (high accuracy mapping of image plane coordinates to geospatial referents) • Nonorthogonal signal processing • Context-based metadata extraction • Very limited work on use of human observers ("ad hoc community of observers")
Level 1: Object Refinement	• Explicit separation of the correlation and estimation problem • Multiple target tracking using multiple hypothesis tracking (MHT) • Use of ad hoc maneuver models • Object identification dominated by feature-based methods • Pattern recognition using neural networks • Emerging guidelines for selection of correlation algorithms	• Dense target environments • Rapidly maneuvering targets • Complex signal propagation • Codependent sensor observations • Background clutter • Context-based reasoning • Integration of identity and kinematic data • Lack of available training data (for target identification) • No true fusion of image and nonimage data (at the data level)
Level 2: Situation Refinement	• Numerous prototype systems • Dominance of rule-based knowledge-based systems (KBS) • Variations include blackboard systems, logical templating, and case-based reasoning • Emerging use of fuzzy logic and agent-based systems	• Limited operational systems • No experience in scaling up prototypes to operational systems • Very limited cognitive models • Perfunctory test and evaluation against toy problems • No proven technique for knowledge engineering
Level 3: Threat Refinement	• Same as level 2 processing • Limited advisory status • Limited deployment experience • Dominated by ad hoc methods • Doctrine-specific, fragile implementations	• Same as level 2 • Difficult to quantify intent • Models require established enemy doctrine • Difficult to model rapidly evolving situations

JDL Process	*Current and Emerging Practices*	*Limitations and Challenges*
	• Very limited ability to predict the evolution of phenomena (e.g., failure phenomena) • Emerging use of hybrid reasoning involving implicit and explicit information	
Level 4: Process Refinement	• Robust methods for single-sensor systems • Formulations based on operations research • Limited context-based reasoning • Focus on measures of performance (MOP) versus measures of effectiveness (MOE) • Emerging use of auction-based methods from e-commerce applications • Emerging use of agents as proxies for bidding for resources	• Difficult to incorporate mission constraints • Scaling problem when there are many sensors (10N) and adaptive systems • Difficult to optimally use non-commensurate sensors • Very difficult to link human information needs to sensor control
Level 5: Cognitive Refinement	• HCI dominated by the technology of the week • Focus on ergonomic versus cognitive-based design • Numerous graphics-based displays and systems • Advanced, 3-D full immersion and human computer interaction (HCI) are available along with limited haptic interfaces • Initial experiments have been conducted with multimodal sensory interactions including sound, touch, and vision • Initial experiments with agent-based cognitive aids (advisory agents using team-based models)	• Very little research has been performed to understand how human analysts process data and make accurate inferences • Creative HCI is needed to adapt to individual users and to provide mitigation of known cognitive biases and illusions • Very limited work on "crowd-sourcing" of analysis (e.g., using virtual world technologies)

process. It has a lengthy history dating back to the invention of the method of least squares by Gauss and Legendre [43]. Enormous amounts of research have been conducted on target tracking, state estimation, and target identification. Indeed, the bulk of the literature in data fusion focuses on this level of fusion. Target tracking is relatively easy in circumstances in which:

- There is a clear partitioning among the data to allow unambiguous allocation of data to specific targets or entities (namely, there is no problem in associating or assigning data observations to targets or tracks).
- The performance of the sensors/observers is known, including the ability to quantify the reliability, performance, uncertainty, and second-order uncertainty of the observations.
- There are well-defined models to link observational data to the desired state vector (well-known and computable observation models, including environmental effects on the sensors).
- The time variation/evolution of the state vector can be predicted using equations of motion along with predictions of the uncertainty of this state vector model.

Violations of these circumstances involve situations such as complex observing environments (e.g., underwater observation of targets in which the propagation media and biological noise is challenging to model), highly maneuvering targets which cause an inability to model the temporal variation of the state vector, low observation rates compared to the dynamic motion of a target, dense target environments (compared with the sensor resolution) that cause ambiguous observation/target association, stealthy targets, and other factors. Other challenges involve situations in which the very concept of a target/entity becomes amorphous or even nonphysical (e.g., a group of humans or a virtual cybernetwork).

A fundamental problem in level 1 processing is called the assignment problem: how to confidently associate observations with unique targets [44]. The process for assignment involves three basic steps: (1) hypothesis generation—generating one or more hypotheses to explain the "meaning of the data" (e.g., possible assignments or interpretations of data/object association), (2) hypothesis evaluation—quantifying the probability, likelihood, or other measure of the relative value of each alternate hypothesis, and (3) hypothesis selection—the actual selection of a hypothesis and assignment of data to a track or state estimate. Techniques for association and correlation span the range from probabilistic association methods, heuristic combination techniques, linear, nonlinear, and integer programming to random set theoretic methods. An extensive review of association/correlation methods was conducted by a team led by J. Llinas and D. Hall [45, 46]. An interesting discussion of the consequences of failing to solve the assignment problem in target tracking is provided by Blackman and Popoli [47].

Assuming that the assignment problem can be adequately addressed, numerous techniques exist for state estimation and update. Classic filtering techniques such as the Kalman filter [48] have been extended to techniques such as the extended Kalman filter [49], unscented filters [50], and, more recently, particle filters [51]. Research combining the estimation process and the assignment process has included multiple hypothesis tracking (MHT) techniques [47] and probabilistic data association (PDA) methods [52, 53]. Other methods include the use of random set theory [54] developed by I. R. Goodman and R. Mahler.

In addition to estimating the state and characteristics of a target or entity, level 1 fusion, entity refinement (target identification), is concerned with entity identification. It can be argued that target tracking and identification are a coupled problem. Mahler [54], for example, maintains that the estimation of target/entity location and identity can be handled under the single general method of random set theory. However, even if the problem of target positional estimation and identity processing is addressed using separate techniques, this is a coupled problem since knowledge of a target's identity can assist in the assignment problem and knowledge of a target's position and velocity can assist in determining what type or class of object is being observed. The basic challenge in target or entity identification is to use the knowledge of observable physical characteristics (size, shape, spectral features, emissions, and movement) to allow a label or name to be associated with the entity. This is a classic problem in statistical classification and estimation addressed by R. A. Fisher in the early 1900s. Examples of target identification include face recognition, specific emitter identification, automated target recognition, and fault identification.

When there is a clear relationship between readily observable parameters and identity and extensive training data exists (namely, a set of data for which we know the true identity based on the observed features), then target identification is straightforward. For example, it is easy to determine the identity of an individual vehicle if there is an observable license plate number. Similarly, we can determine the identity of individual humans if they provide information such as fingerprints, retinal scan information, and corroborating identification card. On the other hand, if such information does not exist or is unobservable, then identification is challenging or sometimes impossible. This is especially true if we seek to impute motive to a human subject or group of subjects (is an observed person or persons a potential threat?).

Methods for target identification fall into three broad classes:

1. *Explicit methods:* use of explicit knowledge for identification. Examples include the use of knowledge such as physical models to link target identity to observable quantities, the use of parametric or logical templates to map the value of observed parameters, and the use of semantic information to identify targets. Such methods are specific to the particular type of target or class of entities that we seek to identify. For example, we might develop a detailed model to predict the radar cross section (RCS) of a target as a function of the type of target, characteristics of the target

shape and construction, aspect angle with respect to the observing radar, and the effects of the intervening signal propagation media. These methods can be computationally very challenging and require detailed knowledge of each type of target to be identified. Alternatively, we may use explicit semantic information to characterize an entity (e.g., a target has a "target-like size," exhibits "target-like motion," and is within a specified geographical area). The semantic information may be represented using production rules, fuzzy logic, probabilistic notions, or other relationships [55]. The use of semantic information for pattern recognition requires a knowledge elicitation from domain experts to obtain the associated rules, parametric boundaries, decision conditions, or the like.

2. *Implicit methods:* use of information "learned" from the data to identify a target. Examples of implicit methods include neural networks, cluster algorithms, machine-learning methods such as support vector machines, and many others [56–58]. These methods generally involve the observation of features associated with a target (e.g., frequency location and shape of peaks in an observed radar cross section, spectral characteristics of an observed emitter, and size and shape parameters) and linking these feature vectors with known groups or clusters in "feature space." Issues with implicit pattern recognition include the availability of sample training data and the selection of appropriate features that provide good recognition performance (namely, separability in feature space) and that are readily observable under a wide variety of conditions.

3. *Hybrid methods:* hybrid methods combine the use of implicit and explicit techniques [59]. These methods tend to be the most robust, because they allow the combination of information "learned" from observed data and also the use of explicit knowledge from physical models or domain experts.

Level 2, situation refinement, refers to the general process of performing automated reasoning to understand the relationship among entities and their relationship to the environment [60–62]. Steinberg [60] provides an excellent overview of situation and threat assessment including an overview of key techniques. The intent of situation refinement is to understand the meaning of the results of level 1 processing—that is, to perform various types of reasoning including object assessment (analysis, recognition, characterization, and projection), and relationship assessment. This involves reasoning to: (1) aggregate objects into larger scale and more generalized entities (e.g., identification that multiple elements such as emitters, weapon systems, and logistical support equipment may comprise a higher-level military unit), (2) identify coordinated communications or movement among multiple entities, and (3) establish causal, temporal, and functional relationships among entities. A broad range of techniques from the field of artificial intelligence [63] have been applied to situation refinement. Examples of techniques include logical templates [64], rule-based production systems (expert systems) [65], Bayesian belief networks [66], and intelligent agents [67, 68].

While numerous techniques and special reasoning architectures such as blackboard systems have been applied to situational refinement, they tend to be challenging to develop and evaluate, and may be "brittle" in the face of real data. All of the reasoning techniques (e.g., rule-based systems, Bayesian belief nets, and so forth) are dependent on the underlying knowledge base (namely, the set of rules, the structure and nature of the reasoning network, the set of logical templates, sample cases for case-based reasoning), derived from domain experts. The process of obtaining the underlying knowledge base has been termed *knowledge engineering* [69], which includes aspects of knowledge elicitation (how to obtain information from the domain experts via interviews, observation, and use of sample problems) and knowledge representation. Examples of knowledge representation techniques include cases, stories, rules, graphs, and others. A continuing issue in knowledge elicitation involves the fact that some domain experts (e.g., situation analysts) may not be able to articulate their train of reasoning or thought in performing their analysis function. The knowledge elicitation process may require assisting them to become aware of precisely how they perform their reasoning. Perusch and McNeese [70] have developed a novel approach they call fuzzy cognitive mapping to help visualize the analysis process. Ultimately, level-2 reasoning seeks to emulate the type of analysis performed by human analysts, and hence will continue to remain a challenge.

Level 3, consequence refinement, processing is analogous to level 2 processing, but seeks to examine current situations and understand how they may evolve in the near future and what the consequences of such an evolution might be. For military applications this involves threat assessment, course of action analysis, and preparation of the battle-space. All of the techniques and comments made for level 2 processing also apply to level 3 processing. A key challenge is to identify plausible hypotheses or potential courses of action and to evaluate them in a timely way to guide decision-making. Figure 1.2 shows the conceptual relationship among level 1, level 2, and level 3 data fusion processes. The combination of level 2 and level 3 processing provides an interpretation of the current situation (e.g., identification and understanding of objects, groups, events, and activities), a prediction of future courses of action, and assessment of three perspectives: (1) the blue view—the location, capabilities, mission, opportunities, and potential actions of our own forces (the "blue" forces), (2) the red view—how the opposing forces view their own capabilities, mission, opportunities, and potential actions, and finally, (3) the white view—how the environment, such as weather, and terrain affect both red and blue.

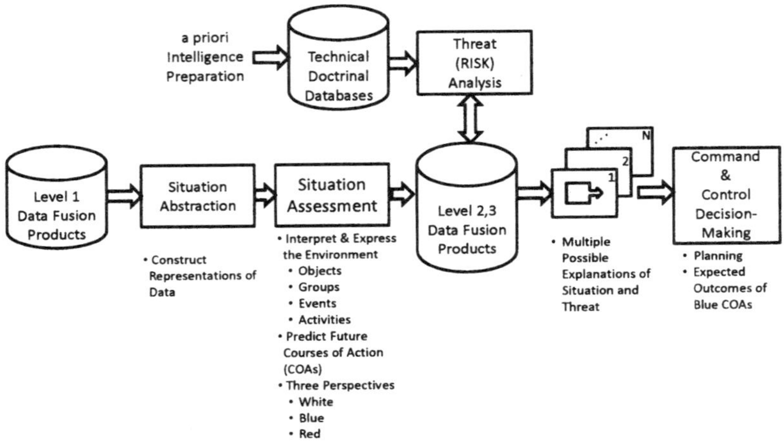

Figure 1.2 Relationship among level 2 and level 3 fusion.

While increasing computational abilities have improved our ability to accurately model physical phenomena such as movement of vehicles, performance of sensors, and effects of terrain and weather, the lack of a known doctrine in asymmetric operations causes a significant challenge. As we will discuss throughout this book, there is a need to model human phenomena as well as physical phenomena.

Level 4, process refinement, processing is a metaprocess; it is a process that seeks to optimize the overall performance of a fusion system. This may entail deployment and utilization of sensors, use of communications infrastructure, adjustment and dynamic selection of fusion algorithms, and interaction among distributed fusions nodes. This is a classic problem in optimization. An enormous amount of literature exists in this area. Methods applied to process refinement for fusion systems have included the use of heuristics, expert systems, utility theory, automated control theory, automated cognition techniques, decision theoretic approaches, probability theory, stochastic dynamic programming, linear programming, neural networks, genetic algorithms, and information theory.

An overview of the overall level 4 problem and potential solutions can be found in Waltz and Llinas [11], Hall and McMullen [1], Mullen et al. [71], and Avasarala et al. [72].

Techniques for optimizing fusion system performance may be broadly grouped into two general categories: (1) control system optimization methods and (2) market-based methods. Control system optimization methods focus on developing a quantitative objective function (or group of objective functions) that describe system performance (e.g., an overall measure of resource utilization, service performance measures, description of factors such as probability of target identification, accuracy of target tracking, surveillance coverage, or many other factors). These are related to potential control parameters such as sensor controls or algorithms controls. Then numerical optimization methods such as linear programming, goal programming, and stochastic dynamic programming are applied to vary the control parameters to achieve the sought-after optimal performance. These methods have a very extensive theoretical basis in control theory and are relatively mature. Challenges include how to quantify the desired system performance, especially in an operational environment of multiple users and varying context of operations. Utility theory and information theory formulations are frequently used in this approach. A second class of methods poses the optimization problem as a dynamic market—treating the sensors and communication system as suppliers of services while fusion algorithms and human users are treated as consumers. Methods from electronic auction theory have been applied for this approach [72, 73]. These methods appear especially suited for situations involving distributed sensors and multiple fusion nodes, with dynamic changes in resources (such as the engagement and disengagement of sources such as human observers). Rapid changes in sensors with improved multifunction performance, increased network connectivity of sensors, information sources, human users, and improved flexibility of system architectures via service-oriented architecture concepts provide increasing opportunities for advanced level 4 processing. At the same time, these very factors make the task of overall system optimization increasingly challenging.

The last level of fusion, level 5, cognitive refinement, seeks to improve the interaction between a fusion system and one or more human users. Level 5 fusion [12] involves techniques in designing effective human-computer interfaces (HCI) as well as techniques to support distributed collaboration among analysts, decision support tools, cognitive bias remediation, and (human) multisensory interaction such as visualization and sonification [6]. We will explore this area in detail in subsequent chapters of this book. Briefly, the mechanics of visualization and computer-human-computer interaction has progressed rapidly, with three dimensional (3-D) full immersion visualization (see Chapter 6), 3-D sound interfaces, haptic (touch-based interaction), and even experimental interfaces based on interpretation of brain waves (a "mind-reading" computer interface) [74]. This area is rapidly driven by the commercial world of virtual world tools and computer games. Collaborative environments such as Second Life, OLIVE, ProtoSphere, and others provide mechanisms for a collaborative environment in a virtual computer world. Computer games such as *Madden NFL* provide external interfaces to accept live weather feeds into the computer games. This provides opportunities to improve the realism of the game environment. Web-based environments such as Whyville (http://www.whyville.net/smmk.nice) provide opportunities for children to practice using virtual world collaboration for educational purposes.

By contrast, tools to support decision-making, focus of human attention, and bias remediation are still emerging. Yen et al. [68, 75, 76] have experimented with the use of intelligent agents based on the recognition-primed decision (RPD) process to improve decision-making, while Saab [76] has explored the use of "cultural lenses" to assist multiple users in understanding each other's perspectives for data interaction. Much work remains in this area to explore what types of techniques are useful and improve collaboration and decision-making.

1.3 NEW ROLES FOR HUMANS IN DATA FUSION

While traditional data fusion systems focused primarily on observation of physical targets by physical sensors, evolving applications are moving towards the characterization of nonphysical targets such as small groups, organizations, and cyberattackers. The focus of traditional data fusion systems is expanding beyond the use of physical sensors to observe, locate, and characterize physical targets (observation and characterization of the physical landscape) to the combined use of physical and nonphysical sensors to observe the human landscape: individuals, groups, populations, organizations, and their interactions. In addition to refocusing from the physical to the human landscape, there is increasing interest in the utilization of nonphysical sensors including humans acting as observers (humans as "soft sensors") and the use of other types of data available on the Web. Figure 1.3 shows these emerging concepts, including: (1) the use of an ad hoc community of observers as a dynamic network of soft sensors to gather information about an emerging situation or activity, (2) the use of human analysts in a hybrid computing mode in which the human's visual and aural pattern recognition skills and semantic reasoning capability are effectively engaged by the information fusion system to assist in the overall situation assessment and decision-making, and (3) the use of a dynamic ad hoc community of analysts (i.e., analytical crowdsourcing) to collaboratively analyze a situation or threat—the interaction may be mediated by a virtual world environment such as Second Life or ProtoSphere. Each of these concepts is summarized next and described in detail in this book.

1.3.1 The Changing Landscape

It is clear that for applications ranging from asymmetric warfare to emergency crisis management to business applications, a need exists to characterize and

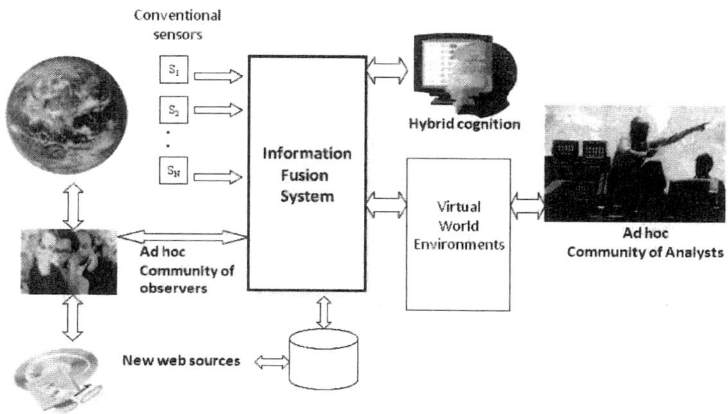

Figure 1.3 Concept of new roles for humans in data fusion systems.

understand the human landscape. In understanding and addressing natural disasters such as Hurricane Katrina in New Orleans [77], it is clearly insufficient to observe and predict weather patterns, model the interaction between high winds and buildings, and so forth. Instead, we must also address information such as population locations and demographics as well as understand how people of different ages, socioeconomic backgrounds, cultures, and experiences might react to a major disaster and to each other. Information and models are needed to address attitudes about the disaster, reaction patterns, reactions to outside agencies and people performing aid, how news media affects the dynamics of the interactions, and many other factors.

On one hand, extensive tools and displays exist to characterize the physical landscape via displays of terrain, weather phenomena, political boundaries, man-made features, and a wide variety of other characteristics. The rapidly evolving Google Earth provides extensive capabilities to obtain information, such as global satellite views of any place on earth, street level views, and local photographs of selected cities and towns, information about streets, maps, distance calculations, Global Positioning Satellite (GPS) location data, and information about businesses. Google has recently arranged to purchase imagery from the GeoEye-1 satellite providing a resolution of 50 cm. Ground-based data collection by Google provides photographs of individual buildings and local features. Similarly, resources such as AccuWeather provide satellite imagery and radar data on the weather throughout the world including 10-day forecasts for selected areas.

By contrast, similar data is not so readily available for analyzing the human landscape. At global and country levels, resources exist to provide information about populations and large organizations and long-term trends. Examples of such sources include:

- Opinion data collected by the Gallup organization via their world database project (http://www.gallup.com/consulting/worldpoll/24046/about.aspx), which collects data on 140 different countries.
- Data collected by the United Nations (http://data.un.org/ and http://www.fao.org/).
- Data from the U.S. State Department (http://www.state.gov).
- The CIA *World Factbook* (https://www.cia.gov/library/publications/the-world-factbook).
- Numerous news organizations throughout the world.

The information collected about the human landscape varies from large-scale, long-term information in the form of facts, population statistics, and political information to minute-by-minute news reports from the news media. Simultaneously, information is being collected in an ad hoc way by humans acting as reporters, bloggers, amateur photographers, and analysts. However, standard geographical information system (GIS) displays and situation displays have not been developed analogous to our routine treatment of the physical landscape via Google Earth or portable GPS devices such as travel aids for automobiles and cell phones. We will explore these issues in the next chapter.

1.3.2 The Human as a Soft Sensor

The rapid growth of cell phone dissemination and continually improving cellular communications bandwidth provide the opportunity to create a dynamic observation resource, in effect allowing humans to act as soft sensors. Information obtained by humans (via direct reports and information from open source information on the Internet) can be valuable and significantly augment data obtained from traditional sensors such as unattended ground sensors, radar, and sensors onboard airborne vehicles. Unfortunately, while extensive techniques exist to combine data from traditional sensors, little work has been done on combining human and nonhuman sensors. Clearly, humans do not act as traditional sensors, and their accuracy, biases, and levels of observation are quite different from traditional sensors. On the other hand, humans can provide valuable inferences and observations not available from standard sensors. A good example is the case in which humans judge that a particular type of relationship exists between some entities. Virtually no hard sensor provides prima facie evidence of the existence of a relationship, since hard sensors are designed primarily to measure attributes and features of entities. A need exists to develop techniques for combining human-supplied data with traditional sensor data. Issues include how to quantify the uncertainty of human data, how to model humans as sensors, how to task humans as sources of information, and even how to elicit information. A summary of some issues related to the use of humans as soft sensors is provided in Table 1.5.

Table 1.5

Issues Related to the Use of Humans as Soft Sensors

Issue	*Description*	*Comments*
Data representation	How should data and information be characterized or reported by a user?	• Humans tend to report information in a self-referential mode (e.g., the automobile is in front of me) and by using fuzzy terminology (terms such as "near" and "far").
Uncertainty representation	How can one characterize the reports provided by a human as a soft sensor? What are the characteristics and performance of human sensors? How are these affected by fatigue, emotion, expectations, and training? In particular, how can one provide metrics for reporting uncertainty and second-order uncertainty?	• Human observers are affected by traditional factors that affect hard sensors (observing conditions, terrain, weather) and also by personal characteristics such as level of training, attention, and fatigue.
Tasking	How can or should we task humans for information (e.g., via requests communicated over a cell phone; use of standard data input forms; encouragement of free texting via systems such as Twitter)?	• Unlike physical sensors, humans do not respond to demands for information and are generally an uncontrolled source. • Tasking may involve a priori agreements and training for selected observers or involve ad hoc reporting such as reporting of emergencies or general "gossip."
Knowledge elicitation	What are the specific mechanisms and methods to elicit information from humans? How can one address common biases without "leading" an observer? What is the role of human aided knowledge elicitation (e.g., a 911 emergency operator) versus computer-aided elicitation via structured forms or guided questions?	• Knowledge elicitation is a well-studied area for developing the knowledge base for expert systems; however, issues in dynamic, ad hoc knowledge elicitation require further study.
Fusion with hard sensor data	How can we effectively combine data from traditional physics-based sensors with human reports?	• Challenges exist in data association and correlation.

Issue	*Description*	*Comments*
		• Fusion of hard and soft data requires accurate characterization of the sources (first- and second-order uncertainties).

Data Representation

Most of the data reported by humans will be in the form of language constructs (sentences, phrases, identifications, and judgments). Antony [78] provides examples of human reports of physical targets and illustrates how the reporting involves fuzzy descriptions and mixed Boolean and fuzzy logic reasoning. He has developed some initial models to convert these descriptions into quantitative expressions of target location, characteristics, and identification. Because the data are generally text-based information, fusion processing of such data requires functions such as text parsing, fuzzy decomposition, use of ontological and thesaurus relationships, and semantic level processing. While humans have a very rich language capability [79], the automated interpretation of this language by computer processing remains a challenge. Issues in data representation are coupled with how the data/knowledge is elicited (e.g., via natural language stream-of-consciousness input versus use of structured input templates and restricted vocabularies).

Tasking

A related challenge in the utilization of human reports involves the issue of sensor "tasking." Just as a traditional sensor must be directed to point in a useful direction (via the computation of "look angles" and sensor tasking), human observers may need to be directed or cued to focus their attention on key targets, events, or activities. While conventional sensor networks are composed solely of physical devices such as radar or infrared detectors, new technologies mean that networked data collection must consider novel "sensors" such as humans sending pictures via cell phone or intelligent software agents combing the Web for information. Correspondingly, sensor management architectures must broaden their abilities to both express complex information gathering tradeoffs to users/decision-makers and effectively task fundamentally different types of sensing entities. The architectural challenge is to intelligently task humans, software agents, and sensor networks to best achieve various high-level goals and directives, given limited resources. Using Web services with service level agreements as a common Rosetta stone, adaptive sensor management middleware would compose and decompose task assignments into actionable subtasks. The system must be able to weigh task assignment tradeoffs such as cost, power consumption, and safety, either independently or by presenting meaningful visualizations to users/decision-makers for feedback. Such a system must be able to merge feedback from end users, from sensor performance measures, and from temporal pattern discovery to learn and to adapt its behavior over time.

Knowledge Elicitation

A key element in utilizing humans as reporters is how to effectively elicit knowledge from ad hoc observers. A range of methods has been tried for commercial purposes. Examples range from automatic teller machines (ATMs) which use a structured menu of inputs, to computer automated help desks, which use natural language processing and restricted hierarchical questions to elicit information from users, to call-in centers such as 911 centers, which utilize live human operators to elicit information about emergencies. Other forms of reporting have included stream-of-consciousness reporting of observed information and activities to specific menus and report formats. A number of issues must be considered in knowledge elicitation, including:

- How can one develop an effective human-computer interface for handheld devices—how can one obtain information in a rapid way in potentially stressful and complex environments (e.g., use of templates, prompts, and menus)?
- What methods should be used to prompt for or assess observer confidence (and second-order uncertainty)?
- What information should be collected to characterize the individual observer and his or her state (e.g., experience as an observer, demographic information, level of stress, mood)?
- What methods (if any) should be used to provide effective feedback and direction to the observer?

1.3.3 Hybrid Cognition

A second major new role of humans in information fusion involves acting in a hybrid computing manner, using human visual and aural pattern recognition and semantic reasoning in collaboration with automated processing performed by a computer. One might consider a human (or multiple humans)/computer team working together in a dynamic way to understand an evolving situation or threat. Pinker [79] notes that humans have two powerful natural cognitive abilities: (1) the ability to recognize and reason with language, and (2) the ability to recognize patterns and reason using a kind of visual physics. For example, sitting in a room, it is easy for a human to identify all of the containers that could hold a liquid—despite the fact that these may include water glasses, coffee cups, pots for plants, a kitchen sink, or a bottle. This would be a daunting task for an automated computer process. Even using sophisticated pattern recognition techniques, the variety of possible containers and even the notion of a container would be difficult to encode into a pattern recognition algorithm. Similarly, we can express situations via sentences, descriptions, or even stories about an event, activity, groups of humans, or collection of entities. Again, despite significant advances in automated reasoning via rules, frames, scripts, logical templates, Bayesian belief nets, or other methods, it is challenging for a computer to match the semantic abilities of almost any human.

By contrast, computers are excellent at prodigious numerical calculations such as those associated with differential equations of motion, fluid flow, statistical estimation, or physics-based modeling. Hence, computers can perform calculations and predictions that are not possible for humans. Clearly, information fusion systems should strive to combine the capabilities of humans and computers to create hybrid reasoning systems capable of performing better than either alone. We will address these concepts in Chapters 6 and 7.

1.3.4 Analytical Crowdsourcing

Finally, we believe that humans can perform a major role in information fusion by dynamic, ad hoc collaboration among multiple people. Examples of dynamic collaboration (sometimes termed "crowdsourcing") are described in [7, 80]. Examples cited in [7] include NASA's Clickworker's project (http://clickworkers.arc.nasa.gov/hirise), which involves using volunteers to help annotate high-resolution image data from the NASA Mars Reconnaissance Orbiter spacecraft, the creation of the Linux operating system by a Finnish computer science student who enlisted the virtual aid of thousands of programmers around the world, and annual contests for development of MATLAB processing scripts to solve challenging problems. Similarly, Sawyer [80] describes collaboration over a period of time, concepts of "group flow" such as group improvisation, customer innovations, and concepts of group genius.

New information technologies such as groupware [81], virtual world tools such as Second Life, ProtoSphere, and OLIVE, multiplayer game environments, and distributed collaboration tools provide enablers for ad hoc collaboration among diverse participants. This interaction may range from an individual consulting a group of experts via a social network, interaction among distributed team members, or an appeal to a large population of potential analysts to support the development of solutions for a problem. Numerous issues involved in such collaborations include how to assess and reward the contributions of virtual team members; how to determine the expertise and capabilities of potential contributors; how to develop an evolving solution that spirals towards a workable solution; how to address issues of proprietary or classified information; and many others. Diverse collaboration is exhibited by wikipedia.com in which numerous contributors are creating an evolving encyclopedia of knowledge. Another example is Amazon.com, which is developing a creative catalog for books, along with information learned by the system such as reader preferences, ad hoc book reviewers, and development of recommendations based on linkages and separate purchases by buyers. Information markets are another mechanism for using large sets of participants to develop predictions ranging from who will win an election to what is a realistic project schedule. Users interact using a betting type of format to specify a solution (e.g., answer to a posed question) and quantify the certainty of their response via bets (usually with virtual money).

While new Web-based technology provides the mechanisms for ready collaboration, the use of these concepts in information fusion to support situation assessment is still in its infancy.

1.4 SUMMARY

While the traditional focus of data fusion systems has been the use of physical sensors to observe physical targets to understand the physical landscape, rapid changes in information technology and changes in focus of interest have motivated a new human-centered view of information fusion. Key changes include:

- Observing the human landscape—changing the domain of interest from the physical landscape to observing and characterizing the human landscape;
- Soft sensors—augmenting the use of physical sensors with humans acting as soft sensors;
- Hybrid computing—use of human-in-the-loop analysis in which humans use their visual and aural pattern recognition capabilities for analysis of complex data and situations, along with semantic reasoning abilities;
- Crowdsourcing of analysis—use of dynamic collaboration among multiple people to support analysis (e.g., via virtual world collaboration).

These new concepts are explored in this book. Chapter 1 provides an introduction to data fusion and the changing environment, while Chapter 2 describes the new human landscape domain of interest. Chapters 3 through 5 introduce the concept of humans as soft sensors, including individuals acting as observers (Chapter 3), groups or communities of observers (Chapter 4), and information available on the Web (Chapter 5). The concept of hybrid computing is discussed in Chapters 6 and 7. Chapter 6 describes the use of advanced visualization for pattern recognition, and Chapter 7 discusses the use of sound (sonification) for understanding data. Chapter 8 discusses the concept of intelligent preparation of the battlefield (IPB) and how this methodology can be adapted to understanding the human landscape or terrain. The last part of the book focuses on the concept of a community of analysts—crowdsourcing of analysis. In particular, Chapter 9 describes the overall concept while Chapter 10 introduces the use of virtual world environments, such as Second Life, for collaborative analysis. Chapter 11 discusses the area of information markets, and finally, Chapter 12 provides perspectives on the future of hybrid data fusion and analysis.

References

[1] Hall, D. L., and S. A. H. McMullen, *Mathematical Techniques in Multisensor Data Fusion*, Norwood, MA: Artech House, 2004.

[2] Liggins, M. E., D. L. Hall, and J. Llinas, eds., *Handbook of Multisensor Data Fusion: Theory and Practice*, 2nd ed., Boca Raton, FL: CRC Press, 2008.

[3] Lucky, R., "A Billion Amateurs," *IEEE Spectrum*, November 2007, p. 96.

[4] Burke, J., et al., "Participatory Sensing," *Proceedings of WSW'06 at SenSys'06*, Boulder, CO, October 31, 2006.

[5] Hall, D., et al., "A Framework for Dynamic Hard/Soft Fusion," *Proceedings of Fusion 2008: The International Conference on Information Fusion*, Cologne, Germany, July 2008.

[6] Ballora, M., et al., "Information Through Sound: Understanding Data Through Sonification," *Proceedings of ISEA2008, The 14th International Symposium on Electronic Art,* Singapore, July 25–August 3, 2008, pp. 47–49.

[7] Howe, J., *Crowdsourcing: Why the Power of the Crowd Is Driving the Future of Business*, New York: Crown Business, 2008.

[8] Nichols, M. L., "Survey of Multisensor Data Fusion Systems," chapter 27 in *Handbook of Multisensor Data Fusion: Theory and Practice*, 2nd ed., M. E. Liggins, D. L. Hall, and J. Llinas, (eds.), Boca Raton, FL: CRC Press, 2008, pp. 691–699.

[9] Kessler, O., et al., *Functional Description of the Data Fusion Process*, Office of Naval Technology Data Fusion Development Strategy, Naval Air Development Center, November 1991.

[10] Steinberg, A. N., C. L. Bowman, and F. E. White, "Revisions to the JDL Model," *Joint HATO/IRIS Conference Proceedings*, Quebec, October 1998.

[11] Waltz, E., and J. Llinas, *Multisensor Data Fusion*, Norwood, MA: Artech House, 2000.

[12] Hall, M. J., S. A. Hall, and T. Tate, "Removing the HCI Bottleneck: How the Human Computer Interface (HCI) Affects the Performance of Data Fusion Systems," *Proceedings of the 2000 MSS National Symposium on Sensor and Data Fusion*, San Diego, CA, June 2000, pp. 89–104.

[13] Blasch, E., and S. Plano, "DFIG Level 5 User Refinement Issues Supporting Situational Assessment Reasoning," *Proceedings of SPIE*, Vol. 4729, 2002.

[14] Boyd, J., *A Discourse on Winning and Losing*, Technical Report, Maxwell, AFB, 1987.

[15] Boyd's OODA Loop, http://www.12manage.com/methods_boyd_ooda_loop.html.

[16] Endsley, M. R., B. Bolte, and D. G. Jones, *Designing for Situation Awareness: An Approach to User-Centered Design*, New York: Taylor and Francis, 2003.

[17] Bedworth, M., and J. O. Obrien, "The Omnibus Model: A New Model of Data Fusion?" *Proceedings of the 2nd International Conference on Information Fusion,* 1999.

[18] Kessler, O., and B. Fabien, *Estimation and ISR Process Integration,* Report for the Defense Advanced Projects Research Agency (DARPA), Washington, D.C., 2001.

[19] Hall, D. L., and R. Linn, "A Taxonomy of Algorithms for Multisensor Data Fusion," *Technical Proceedings of the 1990 Tri-Service Data Fusion Symposium*, Applied Physics Laboratory, Johns Hopkins University, Laurel, MD, Vol. 1, DFS-90, March 1990, pp. 593–610.

[20] Waltz, E., and D. L. Hall, "Requirements Derivation for Data Fusion Systems, " chapter 21 in *Handbook of Multisensor Data Fusion: Theory and Practice*, 2nd ed., M. E. Liggins, D. L. Hall, and J. Llinas (eds.), Boca Raton, FL: CRC Press, 2008, pp. 549–560.

[21] Bowman, C. L., and A. N. Steinberg, "Systems Engineering Approach for Implementing Data Fusion Systems," chapter 22 in *Handbook of Multisensor Data Fusion: Theory and Practice*, 2nd ed., M. E. Liggins, D. L. Hall, and J. Llinas, (eds.), Boca Raton, FL: CRC Press, 2008, pp. 561–596.

[22] Antony, R., *Principles of Data Fusion Automation*, Norwood, MA: Artech House, 1995.

[23] Llinas, J., et al., "Studies and Analyses Within Project Correlation: An In-depth Assessment of Correlation Problems and Solution Techniques," *Proceedings of the 9th National Symposium on Sensor Fusion*, March 12–14, 1996, pp. 171–188.

[24] Llinas, J., "Assessing the Performance of Multisensor Fusion Processes," chapter 25 in *Handbook of Multisensor Data Fusion: Theory and Practice*, 2nd edition, M. E. Liggins, D. L. Hall, and J. Llinas (eds.), Boca Raton, FL: CRC Press, 2008, pp. 655– 675.

[25] McMullen, S. A. H., R. R. Sherry, and S. Miglani, "Survey of COTS Software for Multisensor Data Fusion," chapter 26 in *Handbook of Multisensor Data Fusion: Theory and Practice*, 2nd ed., M. E. Liggins, D. L. Hall, and J. Llinas (eds.), Boca Raton, FL: CRC Press, 2008, pp. 677–689.

[26] Hall, D. L., *Lectures in Multisensor Data Fusion and Target Tracking*, CD-ROM, Norwood, MA: Artech House, 2001.

[27] Srimathveeravalli, G., N. Subramanian, and T. Kesaradas, "A Scenario Generation Tool for DDF Simulation Testbeds," *Proceedings of the Winter 2004 Simulation Conference*, Vol. 2, December 5–8, 2004, *IEEE*, pp. 1698–1705.

[28] Hall, D. L., R. J. Linn, and J. Llinas, "A Survey of Data Fusion Systems," *Proceedings of the SPIE Conference on Data Structures and Target Classification*, Orlando, FL, Vol. 147B, April 1991, pp. 13–29.

[29] Hall, D. L., et al., "Assessing the JDL Model: A Survey and Analysis of Decision and Cognitive Process Models and Comparison with the JDL Model," *Proceedings of the National Symposium on Sensor Data Fusion*, June 2006.

[30] Dasarathy, B., *Decision Fusion*, New York: IEEE Computer Society Press, 1994.

[31] Klein, G. A., *Sources of Power: How People Make Decisions*, Cambridge, MA: MIT Press, 1999.

[32] Klein, G. A., and C. E. Zsambok, (eds.), *Naturalistic Decision Making*, Mahwah, NJ: Lawrence Erlbaum Associates, 1997.

[33] Brehmer, B., "The Dynamic OODA Loop: Amalgamating Boyd's OODA Loop and the Cybernetic Approach to Command and Control," *Proceedings of the 10th International Command and Control Research Technology Symposium,* 2005.

[34] Bryant, D. J., "Rethinking OODA: Toward a Modern Cognitive Framework of Command Decision Making," *Military Psychology,* Vol. 18, No. 3, 2006, p. 183.

[35] Kaempf, G. L., et al., "Decision Making in Complex Naval Command-and-Control Environments," *Human Factors,* Vol. 38, No. 2.

[36] Rousseau, R., and R. Breton, "The M-OODA: A Model Incorporating Control Functions and Teamwork in the OODA Loop," *Proceedings of the 2004 Command and Control Research Technology Symposium,* 2004.

[37] Grant, T., "Unifying Planning and Control Using an OODA-Based Architecture," *Proceedings of the 2005 Annual Research Conference of the South African Institute of Computer Scientists and Information Technologists on IT Research in Developing Countries*, 2005, pp.159–170.

[38] Swanson, D. C., *Signal Processing for Intelligent Sensor Systems*, New York: Marcel Dekker, 2000.

[39] Li, J., and J. Z. Wang, "Automatic Linguistic Indexing of Pictures by a Statistical Modeling Approach, "*IEEE Transactions on Pattern Analysis and Machine Intelligence*, Vol. 25, No. 9, 2003, pp. 1075–1088.

[40] Datta, R., et al., "Image Retrieval: Ideas, Influences, and Trends of the New Age," *ACM Computing Surveys*, Vol. 39, No. 65, 2007.

[41] Goodman, I. R., R. P. Mahler, and H. T. Nguyen, *Mathematics of Data Fusion*, Boston, MA: Kluwer Academic Publishers, 1997.

[42] Brooks, R., and L. Grewe, "Data Registration," chapter 6 in *Handbook of Multisensor Data Fusion: Theory and Practice*, 2nd ed., M. E. Liggins, D. L. Hall, and J. Llinas (eds.), Boca Raton, FL: CRC Press, 2008.

[43] Sorenson, H. W., "Least-Squares Estimation: From Gauss to Kalman," *IEEE Spectrum*, July 1970, pp. 63–28.

[44] Uhlmann, J. K., "Algorithms for Multiple-Target Tracking," *American Scientist*, Vol. 80, No. 2, 1992, pp. 128–141.

[45] Hall, D., and J. Llinas, *Engineering Guidelines for Data Correlation Algorithm Characterization* (SEDI-96-00233), TENCAP SEDI Contractor Report, Lockheed Martin, Contract F04701-95-C-00233, CDRL A0013, July 19, 1996.

[46] Llinas, J., et al., "Studies and Analyses Within Project Correlation: An In-Depth Assessment of Correlation Problems and Solution Techniques," chapter 23 in *Handbook of Multisensor Data Fusion,* 2nd ed., M. E. Liggins, D. L. Hall, and J. Llinas (eds.), Boca Raton, FL: CRC Press, 2008, pp. 597–618.

[47] Blackman, S., and R. Popoli, *Design and Analysis of Modern Tracking Systems*, Norwood, MA: Artech House, 1999.

[48] Kalman, R. E., "A New Approach to Linear Filtering and Prediction Problems," *Transactions of ASME, Journal of Basic Engineering*, Vol. 82, 1960, pp. 34–35.

[49] Gelb, A., *Applied Optimal Estimation*, Cambridge, MA: MIT Press, 1974.

[50] Wan, E. A., and R. Van der Merwe, "The Unscented Kalman Filter for Nonlinear Estimation," *Proceedings of Symposium 2000 on Adaptive Systems for Signal Processing, Communication and Control (AS-SPCC)*, 2000.

[51] Ristic, B., S. Arulampalam, and N. Gordan, *Beyond the Kalman Filter: Particle Filters for Tracking Applications*, Norwood, MA: Artech House, 2004.

[52] Bar-Shalom, Y., (ed.), *Multitarget-Multisensor Tracking: Advanced Applications*, Vol. I, Norwood, MA: Artech House, 1990.

[53] Bar-Shalom, Y., (ed.), *Multitarget-Multisensor Tracking: Advanced Applications*, Vol. II, Norwood, MA: Artech House, 1992.

[54] Mahler, R. P. S., *Statistical Multisource-Multitarget Information Fusion*, Norwood, MA: Artech House, 2007.

[55] De Mantaras, R. L., *Approximate Reasoning Models*, New York: John Wiley and Sons, 1990.

[56] Duda, R., P. Hart, and D. Storke, *Pattern Classification*, New York: John Wiley and Sons, 2001.

[57] Fukunaga, K., *Introduction to Statistical Pattern Recognition*, 2nd edition, New York: Academic Press, 1990.

[58] Bishop, C., *Pattern Recognition and Machine Learning*, New York: Springer, 2007.

[59] Hall, D. L., A. K. Garga, and J. Stover, "Machinery Fault Classification: The Case for a Hybrid Fuzzy Logic Approach," *Proceedings of the 53rd Meeting of the Society for Machinery Failure Prevention Technology*, Virginia Beach, Vol. 53, April 1999, pp. 241–252.

[60] Steinberg, A., "Foundations of Situation and Threat Assessment," chapter 18 in *Handbook of Multisensor Data Fusion: Theory and Practice,* 2nd ed., M. Liggins, D. Hall, and J. Llinas (eds.), Boca Raton, FL: CRC Press, 2008.

[61] Das, S., *High-Level Data Fusion*, Norwood, MA: Artech House, 2008.

[62] Salerno, J., M. Hinman, and D. Boulware, "Building a Framework for Situation Awareness," *Proceedings of the Seventh International Conference on Information Fusion*, Stockholm, July 2004.

[63] Russell, S., and P. Norvig, *Artificial Intelligence: A Modern Approach*, 3rd edition, Upper Saddle River, NJ: Prentice Hall, 2002.

[64] Hall, D., and R. Linn, "Comments on the Use of Templating Techniques for Multisensor Data Fusion," *Proceedings of the Third Tri-Service Data Fusion Symposium*, Applied Physics Laboratory, Johns Hopkins University, Laurel, MD, Vol. 1, May 1989, pp. 345–354.

[65] Ligeza, A., *Logical Foundations for Rule-Based Systems*, New York: Springer, 2006.

[66] Pearl, J., *Probabilistic Reasoning in Intelligent Systems: Networks of Plausible Reasoning*, San Diego, CA: Morgan Kaufmann, 1988.

[67] Fan, X., S. Sun, and J. Yen, "On Shared Situation Awareness for Supporting Human Decision-Making Teams," *2005 AAAI Spring Symposium on AI Technologies for Homeland Security*, March 21–23, 2005.

[68] Fan, X., et al., "RPD-Enabled Agents Teaming with Humans for Multi-Context Decision Making," *Proceedings of the Fifth International Joint Conference on Autonomous Agents and Multi Agent Systems (AAMAS'06),* May 8–12, 2006.

[69] Kendal, S., and M. Creen, *An Introduction to Knowledge Engineering*, New York: Springer-Verlag, 2007.

[70] Perusich, K., and M. McNeese, "Using Fuzzy Cognitive Maps to Define the Search Space in Problem Solving," in G. Salvendy, M. Smith, and R. Koubek (eds.), *Design of Computing Systems: Cognitive Considerations,* Amsterdam, the Netherlands: Elsevier Science, 1997, pp. 805–809.

[71] Mullen, T., V. Avasarala, and D. Hall, "Customer-Driven Sensor Management," *IEEE Intelligent Systems: Special Issue on Self-Management Through Self Organization in Information Systems*, March/April 2006, pp. 41–49.

[72] Avasarala, V., et al., "MASM: A Market Architecture for Sensor Management in Distributed Networks," *Proceedings of the SPIE Defense and Security Symposium*, Orlando, FL, April 12–16, 2005.

[73] Burgoon, J., and V. V. Varadan, "Workshop Report on Detecting and Countering IEDs and Related Threats," NSF Report, http://www.scribd.com/doc/998832/National-Science-Foundation-NSF-IED-Workshop-Report.

[74] Skinner, C. A., "Scientists Build Mind-Reading Computer," *Computer World*, November 5, 2008, p. 16.

[75] Yen, J., X. Fan, S. Sun, M. McNeese, and D. Hall, "Supporting Anti-Terrorist Analyst Teams Using Agents with Shared RPD Process," *Proceedings of the 2004 IEEE International Conference, Computational Intelligence for Homeland Security and Personal Safety, CIHSPS 2004*, pp. 53–60.

[76] Saab, D., "Ontological Complexity and Human Culture," *Proceedings of the Conference on Philosophy's Relevance in Information Science*, Paderborn, Germany, Oct 3–4, 2009.

[77] Palser, B., *Hurricane Katrina: Aftermath of Disaster*, Minneapolis, MN: Compass Point Books, 2005.

[78] Antony, R., "Data Management Support to Tactical Data Fusion," chapter 24 in *Handbook of Multisensor Data Fusion: Theory and Practice*, 2nd ed., M. E. Liggins, D. L. Hall, and J. Llinas (eds.), Boca Raton, FL: CRC Press, 2008, pp. 619–653.

[79] Pinker, S., *How the Mind Works*, New York: Norton, 1997.

[80] Sawyer, K., *Group Genius: The Creative Power of Collaboration*, New York: Basic Books, 2007.

[81] Schummer, T., and S. Lukosch, *Patterns for Computer-Aided Interaction*, New York: Wiley, 2007.

Chapter 2

Sensing the Human Landscape: Issues and Opportunities

As humans connect to each other in more powerful and more widely available ways, we witness an amazing array of consequences, many of them unexpected. Small, coordinated groups—"smart mobs" in Howard Rheingold's phrase—can make an impact on culture, economics, and security [1]. At the same time, the realities of armed conflict in the twenty-first century increase the need for understanding of indigenous populations with often unique social, cultural, and political dynamics. Thus, we are witnessing in many sectors—commercial, military, and philanthropic/nongovernmental—an increased awareness of the need for systematic, timely, and reliable information regarding the human landscape.

2.1 INTRODUCTION

While social science has long taken as its mandate the study of human populations in their economic, historical, demographic, and other dimensions, until recently the discipline of information fusion has accounted for people in only selected ways. Because of the changing nature of conflict, communications, and constituencies, however, the human landscape has emerged in the past several decades as a key element of understanding for analysts and decision-makers alike. The concept of *human landscapes* accounts for many dimensions of human activity at the aggregate level.

2.1.1 Definitions

Traditionally, humans have been considered distinct from the landscape. In fact, the *Oxford English Dictionary* defines landscape as "inland natural scenery, or its representation in painting." From the American perspective, *Webster's Second New International Dictionary* takes a similar angle: "a portion of land or territory which the eye can comprehend in a single view, including all the objects so seen, especially in its pictorial aspect." The emphasis, then, falls on broad swaths of terrain, seen from a single perspective, ultimately for the purpose of representation. The representations can, in turn, support many different processes of cognition, hypothesis-testing, or decision-making.

In Figure 2.1, we indicate that the traditional view of information fusion systems has focused on the physical landscape (including physical objects, terrain, weather), with some information overlaid over human-made objects such as buildings and roads, and human-defined concepts such as political boundaries; in general, there has not been a focus on the human landscape.

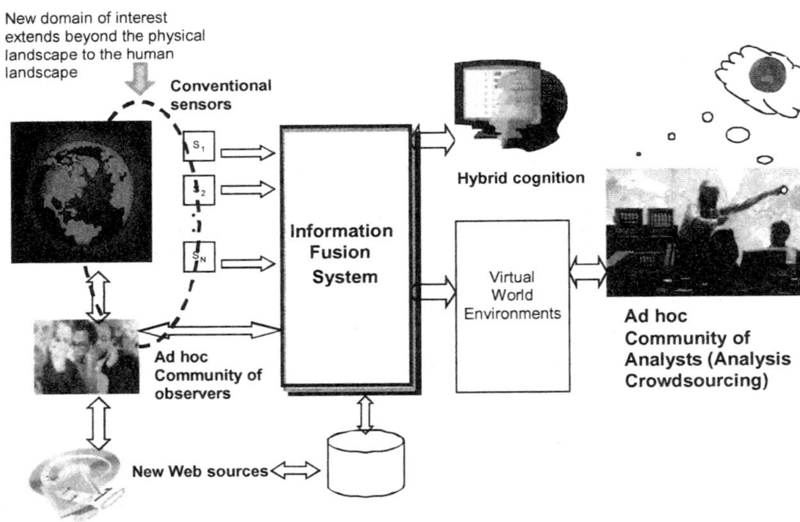

Figure 2.1 New focus on the human landscape. Beyond the traditional information fusion of data associated with the physical landscape.

For our purposes, the human landscape might be understood as people, in the aggregate at various scales, as they relate to the natural world, the built environment, and each other. Note, we have deliberately introduced the term "human landscape" rather than use the term "human terrain." *Human terrain* is a term dominant in the U.S. Department of Defense [2] and is defined as the human population and society in the operational environment (area of operations) as defined and characterized by sociocultural, anthropologic, and ethnographic data and other nongeophysical information about the population and society. While the concepts of human terrain and human landscape are very similar, we prefer a more general term that can relate to nonmilitary operations (e.g., business planning, emergency crisis management, environmental monitoring, and other general concepts). Moreover, the term human terrain is also related to military concepts such as the *human terrain system* (a proof-of-concept program run by the U.S. Army Training and Doctrine Command), human terrain teams [3] (which are five-to-nine person teams assigned to brigade combat teams for improving the understanding concerning cultural and social issues in an area of operations), and *human terrain mapping.*

Several important ideas should be mentioned here. The human landscape consists of people in the aggregate rather than individuals, who are more properly the subject of biography, criminology, or intelligence-gathering. People can be aggregated by location (village, state, region, or nation-state), by family or tribe, by ethnic identity, by economic indicators (trade, socioeconomic status, wealth/income), and many other traits. People in groups, in turn, relate to the natural world: the land, air, and sea; the animal kingdom; and to food and vegetation. People also interact with the made environment—cities, modes of transportation such as roads or ports, and economic venues including markets and factories—and with means of communications. Finally, people interact within their aggregated groups, and their groups relate to other groups.

So defined, the human landscape is obviously vast, and unknowable in its entirety. The scale is a product of both the large number of individuals and the much larger number of potential interactions within a group: if 8 people can meet in any of 246 possible combinations, the potential dynamics of a village, much less an online community, quickly outgrow the available mathematical models. Complicating matters further, the human landscape can change much more rapidly than the natural or made environments, so tracking change over time adds a further dimension of complexity. Nevertheless, integrating knowledge relating to the human landscape with information traditionally derived from electromechanical sensors presents opportunities of great value to many constituencies.

We note that while the complications of defining and representing the human landscape appear daunting, similar comments could be made about the physical landscape. For example, if we sought to represent the physical landscape at the highest level of fidelity and completeness, it would imply requiring information about terrain, characteristics of the soil such as acidity, vegetation information, data about small animals and microbes, micro-weather conditions, and many other factors. However, we are used to abstractions that focus our attention on the features of the physical landscape that relate to our purpose or mission at hand. Thus, a representation of the physical landscape for the purposes of travel from one city to another by automobile would require only information about roads, traffic conditions, and estimated travel times, but would not require information about vegetation, soil characteristics (unless we were traveling off road), micro-weather conditions (unless they affected travel conditions), and so on. In short, we are quite used to presentations related to the physical landscape based on maps, GPS devices such as TomTom or Garmin, or Google Earth characterizations and overlays.

2.1.2 Context: Why the Current Interest?

The current interest in the human landscape can be traced to at least four interrelated causes. First, the speed of communications made possible by television, the Internet, and global cellular adoption has truly shrunk the globe. Air travel and immigration bring people in closer contact (electronic and otherwise) than ever before with different groups with various views, objectives, and modes of understanding. At a practical level, facilitating these new kinds of interactions, at scale, requires better methods for representing relevant attributes of different groups. Cultural signals such as who speaks in a meeting, how words and concepts are or are not translated, and how position is represented can be subtle yet matter significantly. Commonly used gestures in Western culture can be taken as insults in non-Western cultures. For example, the "ring gesture" in which the thumb and forefinger are touched together to form a circle has greatly different meanings in the United States, Greece, and Turkey.

Second, advances in the mathematics and science of networks in recent decades are making possible new kinds of models and representations of larger groups of people in their relationship to others [4, 5]. Such commercial services as Facebook, MySpace, and Hi5 are implementing these models at an unprecedented scale, and the state of practice currently outpaces theory. Moreover, Tapscott [6] suggests that the new generation of digital natives is increasingly using social networks in all aspects of their lives.

Third, the rise of nonnation-state political actors, whether the World Bank, Greenpeace, or Al Qaeda, increases the importance for understanding people in groups from multiple perspectives, whether for the purposes of humanitarian assistance, accountability, or security. The need for understanding the economic rationale for, and implications of, tribal conflict and subsequent Philippine deforestation, for example, may unite entities as disparate as the Red Cross/Red Crescent, the International Monetary Fund, the U.S. State Department, and various armed forces.

Finally, international crime and terrorism are based in human networks much more than in formal political structures, physical assets, and geographically delimited space. Because multiple countries, agencies, and kinds of agencies are enlisted in the effort to prevent and counter the rise of these entities, the human landscape has joined a long list of essential elements of military readiness.

2.1.3 Constituencies for Human Landscape Information

Traditionally, information regarding the human landscape fell into the province of intelligence agencies, census and statistical bureaus, and international development groups such as various United Nations, World Bank, and nongovernmental bodies. These groups remain essential sources for many elements of human landscape data: available dietary calories, life expectancy, religious and ethnic affiliation, income inequality, and many more. Recently, however, more groups are interested in the human landscape.

In part because of standardized data reporting made possible by the Internet, many people want to know more about the human landscape because the limits of paper recordkeeping are being both exposed and transcended. That is, people want to know more because the transparency of such services as air freight package tracking, online news sources, stock and other financial information, and countless other examples have served to raise expectations. Statistical agencies in many parts of the world are responding, which only serves to raise expectations further.

Philanthropy is also in a fertile period of innovation. Just as the generation of Carnegie, Mellon, the Rockefellers, and Ford invented the foundations that bear their names, today's philanthropists are reinventing their practices. Bill and Melinda Gates, Google.org, and Omidyar Network (based on the eBay fortune) represent three of many new models for improving the lot of humanity, and to do so, all require more information about the human landscape than currently exists [7, 8].

Between philanthropy and nation-state status fall a wide variety of nongovernmental organizations (NGOs) that address a range of issues that almost always, at least indirectly, touch the human landscape. Whether it is whaling or disaster relief or economic development, NGOs are emerging as both consumers and producers of data related to the human landscape.

Economic development is also merging with the world of commerce as practices such as micro-lending—small, targeted loans that both spur economic growth and return profits [9]—demand new types and quality of information. Traditional marketing practices translate poorly to places such as Africa. New organizational forms are emerging to challenge Western models of capitalism, whether an Asian trading company, a Korean *chaebol,* or even Wikipedia. Word of mouth in every market is amplified by today's communications and media networks [10]. Complex global logistics processes can be disrupted by such factors as avian flu, Somali piracy, or Mexican narco-terror. In each instance, the world of business joins the list of constituencies for improved management of data related to human landscapes, not only in the economic dimension.

Finally, as we have seen, asymmetric warfare is challenging conventional forces to comprehend and counter "soft" assets as well as large-scale armies and weapons systems. The Mumbai terror attacks of November 2008, for example, were facilitated by satellite phones at sea, GPS receivers, voice-over Internet Protocol communications (which was untraceable with conventional telecommunications tools), satellite imagery, and real-time communications between handlers watching live television coverage and terrorists on the ground [11]. All of these technologies, along with training, ideological commitment, and the element of surprise, made them a small but lethal force. Given that the ready availability of the tools cannot be constrained, understanding the human networks of terror has become a high priority for security forces worldwide.

2.2 CONTRASTS

Human landscapes can be understood as extensions of the natural landscape: such features as water, crops, and mountains constrain what people can do and where they can go. At the same time, geospatial characteristics often fail to correlate with religious beliefs, opinions, and ways of making a living. A major challenge for understanding the human landscape lies in linking spatial and nonspatial dimensions of belief, capability, and organizations.

2.2.1 Similarities Between Natural Terrain and Human Landscapes

The human landscape shares with its natural counterpart a spatial dimension: a given individual can only be in one place at one time, movement from point A to point B can be measured in kilometers, and many human attributes relate to people's geospatial coordinates (farmers cannot farm in cities, fishermen cannot pursue their catch inland, language barriers typically have spatial demarcations). Thus, for many facets of the human landscape, a map can serve as a base layer.

This model means that the familiar layering metaphor from terrain mapping can at times be useful. Instead of rivers, roads, water lines, annual rainfall, or other observed additions to the base layer, human landscape layers might concern diet, religious affiliation, language tendencies, family or tribal affiliation, economic well-being, and so on.

Many natural phenomena are emergent; that is, complex systems and patterns arise out of a multiplicity of relatively simple interactions. Bird flocking, sand dune movement, weather, and cellular phenomena provide illustrations. Similarly, many features of the human landscape are emergent: traffic jams are a classic example, but soccer riots, financial markets, and Internet traffic all follow parallel paths from simple rules to complex behavior. In both domains, identifying the relevant precursors and projecting emergent effects is typically difficult or even impossible: emergent properties are highly nonlinear and fall outside most available models [12].

2.2.2 Differences Between Natural Terrain and Human Landscapes (What Google Maps Cannot Convey)

At the same time, the human landscape presents challenges to a geospatial overlay model. Belief systems, which are often difficult to discern in the first place, often cross natural boundaries. Even more commonly, political borders may have been superimposed on longstanding tribal, ethnic, or economic arrangements. The Kashmir portion of the India/Pakistan/Afghanistan region presents one ready example.

The electronic trails left by digitally connected humans can often be spoofed or disguised: the area code of a voice over IP phone call may or may not reveal the caller's physical location. Unlike traditional wireline phones, cell phone numbers on caller ID tell the recipient nothing about physical presence. Photographs can be easily and subtly manipulated. E-mail addresses are routinely exploited for spamming and other purposes. In many, if not most, instances, the digital evidence relating to a human in the physical or virtual landscape cannot be taken at face value [13].

Many aspects of the human landscape can only be discerned by other humans; identifying someone carrying a concealed firearm is one example. Unlike most situations in physical sensing, the presence of a human observer/analyst can change the phenomenon under observation. In a tribal ritual of human burial, whether in Brazil, New Jersey, or Tokyo, the presence of a stranger in the midst will change what the assembled group will do and how it will do it. In contrast, for much of the natural world, if a rock is being measured for hardness or for the presence of ferrous metals, the observer rarely alters the elemental composition of the rock. Taking an aerial photograph of a river will not move the path of the water.

Similarly, direct interrogation of a natural event is often standard procedure: Has this ground been disturbed? Has this door been opened in the past x hours? Does this body of water exhibit artifacts of submarine traffic? For human landscapes, asking individuals or groups direct questions can violate cultural norms, expose the questioner or respondent to harm or loss of credibility, or generate responses that, while satisfying to the questioner, may fail to reflect underlying realities. This is one of the reasons why political polling practices sometimes produce less accurate results than electronic information markets [14]. The need for tact, diplomacy, guile, and other intangible but valuable traits in human evaluation of human landscapes represents yet another departure from sensors of physical or electronic landscapes, where other intangibles are required for success.

Because human landscapes both adhere to and diverge from geospatial dimensions, tracing cause and effect can be difficult. Two events, geographically distant, could be connected by a kinship network, a long-held grudge, or serendipitous randomness. Knowing the relevant context for sets of events or observations, given often subjective understandings even on the part of the principals involved, is frequently impossible. Thus, many visualization and analysis tools that have proven themselves in scientific computing or GIS, for example, translate only poorly to the many "soft" facets of the human landscape.

2.2.3 The Richness of the Human Landscape

Humans have developed a phenomenal range of systems of expression, of meaning, and of surviving. When looking at one facet of a group, it is often difficult to hold another constant or in tension. Observing how food production and attitudes toward recreation interact, for instance, is seldom a trivial exercise, but the two may be deeply connected, as they were in nineteeth-century America. A short list begins to suggest the range of sensing and sense-making mechanisms required to understand a given population.

Groups

How do people organize themselves? How are they organized by outside powers or authorities? What factors determine insider versus outsider status? What makes another group perceived as friendly or irrelevant or an enemy? What symbols signify identity or lack thereof within a population?

Political Institutions

How is the status of the various groups instantiated into tribal, statutory, and/or case law? How is authority within a group earned, lost, and signified? What political institutions protect the group against both internal (criminal) and external enemies? What institutions perpetuate the identity of the group(s), its laws, processes, and self-understanding? These can be as formal as a document (the Magna Carta) or as informal as Independence Day celebrations.

Survival

How are food, clothing, and shelter provided? What rituals accompany each of these? How does the pursuit of survival map to other social dynamics, such as power and influence? For example, how does housing type and location relate to social status? At a communal table, who eats what? Who eats first, last, or in separate quarters?

Quality of Life

How are individual and public health supported? What are the group's highest and lowest priorities as measured by rhetoric, spending, and trade-offs? How, if at all, do such values as natural beauty, contemplation, and storytelling find expression?

Making a Living

How are work, money, and possessions understood? How much of the economy is regulated and/or taxed, and how much is transacted informally? How does a given group manage its banking and credit processes, if at all? How does the group support or inhibit the ability of its members to trade, profit, or otherwise expand their economic horizons? What is the status of the economic infrastructure, whether legal (contracts), commercial (banks or trade regulators), or physical (ports, telephony, roads)?

Natural Life Cycles

How are birth, childhood, maturation, adulthood, and death understood, managed, and represented? How do value judgments relating to the status of different stages find expression in lore, law, and practice? How is the group's interrelationship with land, water, and climate understood and conducted?

Religion

What is the range of accepted religious belief and practice in a given group? What rituals animate religious doctrine? What is the relation between church/mosque/ synagogue/ashram and the state? How do various authority figures reward conformity and sanction alternative expression?

Gender

How are relationships between the sexes regulated? How are norms and violations of those norms defined? What are the consequences of variation from orthodoxy? How do norms and behaviors vary when men and women are apart compared to when they are commingled?

Recreation and Leisure

Is there sufficient social surplus to allow for relaxation, sport, and pursuit of personal interests such as handcrafts or music and dance? What is the relation between formal political authority and/or group identity and sport, as in Soviet Olympians, Brazilian football, or Indian cricket?

2.3 ELEMENTS OF THE HUMAN LANDSCAPE

The dimensions of measurement of aggregate human characteristics are as varied as people themselves. Any given attribute, whether height or level of education, is both contextual and an indicator in and of itself. Understanding attributes as elements of systems and subsystems begins the process of standardizing the measure of a given population.

2.3.1 Attributes: Chosen Versus Given

When evaluating the human landscape, observers can distinguish between attributes over which an individual has little or no control, such as height or age, and attributes which can be chosen, such as political party affiliation or occupation. While the latter are often more interesting, they are typically more difficult to collect. People routinely lie, obscure answers, or change their mind after speaking to a pollster or focus group coordinator, for example. Furthermore, technology has, over time, increased the number of things people can choose: hair color or coverage can now be a matter of discretionary income as much as heredity. Eye color is no longer assumed to be given. Other attributes, while not inherent, are difficult to fake: calluses, gait, and voice and accent fall into a middle ground of often unintentionally revealing attributes that nevertheless cannot be taken at face value.

Furthermore, chosen attributes may be chosen largely for deceptive purposes. When dealing with subtle matters such as kinship, age, affiliation, occupation, income, and other revealing attributes, the observer or questioner often can do little but rely on the responses of his or her contacts. Taking everything at face value is rarely an option, but neither is questioning every response. Finding appropriate middle ground is difficult.

Representing chosen attributes is accordingly more difficult than capturing given ones: saying that individuals, or members of a group, are of a certain height fits traditional data models quite comfortably. Trying to elicit and then represent attitudes toward change or personal security, or the "fashion sense" of a group as expressed by their attire [15], tests quantitative metrics and thus traditional fusion methods. Generally speaking, chosen attributes are softer and thus more difficult to manage at all stages of an information life cycle.

2.3.2 Attributes: Sensed by Humans Versus Sensed by Technology

Technology can discern many elements of the human landscape, presence foremost among them. Whether by cell phone signals, thermal radiation, photography at various spectra, or sensing mass, technology is extremely useful in telling us that a body, and sometimes which body, is in a given space. Technology can sometimes detect what people are carrying or wearing (especially at airports), what they ingested (as the presence of drug residue in the public sewage system would reveal), or what they buy through point-of-sale scan data.

At the same time, many important human attributes can only be discerned by human beings, sometimes highly trained ones. The security practices of Israel's El Al national airline, for example, rely far more heavily on human interaction than on technological solutions, with impressive results but significant ramifications for other kinds of societies and cultures [16]. Machines do a poor job understanding figures of speech, as automated translation programs show. Only a person can ask "what did you mean by that?" after hearing a statement or observing an action.

2.3.3 Attributes: Individual Versus Collective

Some beliefs embraced or actions taken by humans are indicative only of the individual concerned. Knowing when a person speaks for herself and when she represents a larger body of thought or experience is difficult to ascertain. At the same time, individual expressions can be powerful lenses into broader sentiment. How can the observer differentiate?

Traditional social science methodology relies on statistical sampling techniques. Presidential polls and television ratings, for example, gain their authority from the size, composition, and other attributes of the sampled population. The limits of sampling are well known; however, a representative sample of toothpaste users might not work for determining automobile preference or attitudes toward capital punishment. In addition, as the science of networks, human and otherwise, builds on the insights of Erdos, Milgram, and others, to date there is no reliable technique for generating statistically reliable samples of human social networks [5].

2.3.4 Attributes as Seen by Various Disciplines

Just as a sensor on a satellite or other platform might operate at one of many wavelengths in order to track motorized vehicles in a jungle, propeller aircraft, drought, or schools of fish, so, too, do multiple disciples focus their attention on various facets of the human landscape. As might be expected, these areas of inquiry overlap considerably.

Economics

Economics focuses attention on human behavior as it relates to the production and consumption of goods and services. Scarcity and surplus drive much of economic thought, which is conventionally divided between microeconomics (which focuses on individuals and firms) and macroeconomics (which takes as its subject entire economies). In addition, much of economic thought and practice is concerned with the interplay between incentives and constraints, as alternative uses for a given resource are evaluated and pursued.

Criminology

Criminology studies crime as an individual and social phenomenon. It overlaps with law, sociology, and psychology as its students and practitioners attempt both to understand and prevent crime. This has historically been addressed both through the use of incentives, such as job training or education, and punishment, including incarceration or the death penalty.

Sociology

Sociology studies use a combination of empirical investigation, such as surveys or observation and theoretical analysis, to attempt to understand human social structure and activity. Sociology can be qualitative or quantitative, pursue both pure knowledge and social improvement, and focus on groups of any size from a dyad (pair) up to entire societies.

History

History quite simply studies the past. The particular focus areas of historical inquiry can often provide considerable leverage to an observer seeking to understand a given human landscape. Why does a given group live where it does? What are the traditional and modern ways of making a living? How are ideas related to religion, the supernatural, and a given group's origins understood and perpetuated? How does a group coexist with nearby groups given relevant events and attitudes from prior times? How and why has the social status of different groups such as women, children, or ethnic minorities changed over time? Who are the heroic figures in a group's past, and how has daily life of the often invisible masses interacted with the behaviors of those elites?

Anthropology

Anthropology attempts to study humans and the state of humanity from the broadest possible perspective. The discipline is defined different ways in different places. In the United States, anthropology is broken down into four fields. Physical or biological anthropology studies the physical human being through such lenses as human evolution, population genetics, and primatology. Cultural anthropology studies culture and social organization, looking at economics, law, religion, family structure, gender relations, and the like. Linguistic anthropology focuses on processes of nonverbal and verbal communications, while archaeology looks at the artifacts of a culture, overlapping heavily with the work of historians.

Psychology

Psychology can be either an applied (clinical) or academic study of human mental functions and behavior. The field overlaps heavily with medicine, law and criminology, sociology, biology, anthropology, economics, and other disciplines. Both the unconscious and conscious mind are studied, the latter in its activities of perception, cognition, personality, behavior, and interpersonal relationships.

Public Health

Public health looks to prevent disease and promote physical well-being in organized groups of individuals: schools, towns, tribes, and subpopulations such as ethnic groups. The focus is primarily on prevention rather than cure, so statistical analysis of emerging issues is a primary tool in the pursuit of education and behavioral change (antismoking campaigns), infrastructure development (water and sanitation), and clinical intervention in the form of inoculation or condom distribution.

Law, journalism, and business scholarship, such as market research, can provide further insight into the human landscape.

2.4 ISSUES IN REPRESENTING THE HUMAN LANDSCAPE

For human-centered fusion, the human landscape presents a particular challenge insofar as people use words in specific yet nonstandard ways, presents both explicit and implicit signals, and can introduce ambiguity both intentionally and unintentionally. Deriving systems of meaningful information from human populations has long taxed individual disciplines; fusing multiple streams of human landscape information presents a challenge far greater than the sum of the various disciplinary parts.

2.4.1 Tacit Versus Explicit Knowledge

Beginning with the philosopher Michael Polanyi in the early 1960s, the varieties of human knowledge have become clearer. His distinction between tacit and explicit knowledge is directly relevant to the task of understanding human landscapes. Polanyi was direct: "We know more than we can tell" [17]. That is, all of us can recognize acquaintances by face or sometimes by voice, but we struggle to tell a third party how to recognize our friend. Similarly, speech accents can be recognized but not well described, and while most Americans can distinguish the accent of someone from South Carolina from that of her cousin visiting from Maine, few can articulate the differences.

Explicit knowledge, in contrast, can be conveyed in writing: "turn left at the third stoplight on Main Street after passing the courthouse on your left" should be unambiguous. Recipes and other directions can occupy a middle ground: a pound of butter is hard to confuse, but calling for a pinch of salt, or dough that is "tacky," can mean different things to different people at different times. *The Fannie Farmer Cookbook*, published in 1896, addressed this issue directly as it was the first U.S. cookbook to use explicit, standardized weights and measures [18].

When representing the human landscape, the differences between tacit and explicit knowledge are of great consequence for the fusion process. The statements "my contact in the village seemed glad to see me" versus "the village constable was wearing a navy waistcoat and a formal brimmed hat" must be understood in importantly different ways that have nothing (necessarily) to do with the truth or falsehood of the statements. In addition, what might be reliably known by an observer might not be reliably interpreted by someone outside the direct context.

2.4.2 Indirect Human Evidence Is Often More Reliable Than Direct Inquiry

Unlike the natural world, the human landscape is difficult to apprehend with direct evidence. A mother-in-law asking her daughter's husband if he likes his birthday gift is likely to produce a polite answer in the affirmative even if the sweater in the gift box never gets worn. Asking people how much confidence they have in the economy will produce verbal evidence often at odds with macro consumer spending patterns. For matters of delicacy or to avoid repercussions, direct questions, particularly from a live interviewer, are likely to produce evidence at odds with behavior. Even after 50 years, Alfred Kinsey's methodology for determining Americans' sexual attitudes and practices remains controversial [19].

Many methods of indirect investigation have proven successful. Water flows during major televised sporting events reveal patterns of television viewership based on bathroom breaks. Gait analysis is more difficult to outwit than other physical surveillance tools. For every person who tells a doctor or pollster that he wants to lose weight, tales of fast food and snack food more directly correlate with the reality of a near epidemic of obesity. Similarly, bond prices reveal mass sentiment toward the prospects for the future. Even though multiple textbooks portray the run-up to World War I as "gathering storm clouds" or a similar figure of speech, bond prices indicated no such concern on a wide basis [20].

2.4.3 Context Is Often Equally Important as Data

Knowing the source of a piece of information serves multiple purposes: it validates or raises concerns about the quality of the data or observation, it situates the information among multiple observations, and it can help guide the use of formal or informal "fudge factors" or other weighting instruments. For both traditional sensor data and information regarding the human landscape, context is also essential for identifying change or stasis over time. When those changes are qualitative, as in public opinion, however, the context can be difficult to convey. Representations of soft attributes are difficult to benchmark or calibrate, particularly as learning effects take hold. That is, as people learn more about a phenomenon, they think of questions they might have asked earlier to avoid dead ends, add nuance, or generate alternative hypotheses. Context for human landscape data can come in a variety of forms, as seen in the following.

Says W*ho?*

Did the observer experience the reported experience first-hand? Who determines what the relevant characteristics are for a given interaction or category of interactions and how they are measured? In essence, issues of meta-information grow rapidly.

When?

If the information is derived from human observation, what are the characteristics of the observer making an observation at time t_1 compared to those attributes of the same observer making an observation at a subsequent time t_2?

Where*?*

Humans move, and human attributes may or may not move with them. A farmer who moves to the city may become a cab driver or a welder, for example. Hearing about some phenomenon might provide insight into local phenomena, but determining just how local can be complicated.

Which?

In many countries, surnames provide little disambiguation. Structurally, political authority might be held by birth order, family position, or intimidation. Determining who is who, then who controls a given village or supports a given policy, often takes long experience and deep connections.

2.4.4 Representation of Time in Human Landscapes

Timescapes vary considerably in human populations. In groups with strong oral histories (sometimes to the exclusion of writing), past generations are included in nearly a present sense. Elsewhere, attitudes and behaviors can change in an instant, as in a soccer riot or the aftermath of a natural disaster. Thus, distinguishing between what is near-term and what might lie in the distant future will vary by issue, by population, and by time. Guessing when Saudi women might be permitted to vote, in the age of text-messaging and rapid global cell phone adoption, is an entirely different matter from predicting the fall of the Berlin Wall before broad adoption of the Internet.

Perceptions of time generally differ between the observer and the population in question. Gauging the proper length of a formal meal in Tokyo, Rome, or Mexico City is difficult without close contact with local populations. One table's "we've been sitting here forever" is the next group's "why did the food come so fast?" When it comes to reporting something with a time dimension, how does the outsider convert between his time sense and the predominant local norm? How are those differences managed in subsequent processing?

Unlike time with physical sensors, which may be driven with extreme accuracy, as with GPS satellites, time in human landscapes is often relative rather than absolute. That is, a town may have learned of a tragedy after sundown, or before sunup, rather than at 2100 hours. This applies to the past as well to the future. Saying that something will occur "after the crops come in" is common in agrarian cultures, but observers attuned to urban rhythms will likely have difficulty with both translating the reference and living with the uncertainty connoted by the remark. Triangulating among multiple events so denoted will likely require new forms of error correction and/or probabilistic representation.

2.4.5 Presence of Humans as Sensors Can Change What They Observe

The concept of a certain social gaffe appears to be nearly universal. Some judgment is rendered on a third party presumed to be absent, and then said person appears from around a corner, having heard everything. Manners apart, humans are selective with what they say about other humans, whether individually or in groups. Whether a given piece of information will be shared with an outside questioner depends on the person speaking, the topic under discussion, and the outsider. Accounting for various possible scenarios regarding each of the three and their interactions once again adds complexity to any findings that may be gathered.

Similarly, the cry of "hey, watch this" indicates that the presence of an observer (whether a mother or a teenage peer) is affecting the behavior of the person riding the cycle or attempting a jump. The observer thus needs to be aware of his or her relation to the human interactions being witnessed and calibrate if they should be discounted, corroborated, or taken at face value.

2.4.6 Representing Uncertainty

In the absence of robust statistical confidence, corroborating verbal evidence, and/or electronic or written evidence, what happens when evidence regarding the human landscape is collected, cleansed, and processed? How does "I think I understood this" get captured in large, formal systems of information fusion? Will a system much like musical notation emerge, in which the notes do not change, but how they are played is influenced by the composer's tempo and other commentary? Is there a place for "soft" confidence intervals to be notated and preserved in processing? Will quality metrics lead to discount factors being applied to a given observation? Given that certainty with regard to human reports about human populations is an impossible goal, the task of reliably, unobtrusively, and consistently representing uncertainty remains as yet another challenge in the development of human landscape information fusion.

2.4.7 Unique Identifiers for Humans Versus Other Entities

The use of unique identifiers for human beings raises multiple issues relating to civil liberties, privacy, and risk. While serialized identifiers are commonplace in industrial, military, and other applications, and they have the potential to eliminate confusion, the issue of standardized drivers' licenses in the United States illustrates some of the perils. State governments and other groups are fighting the Real ID legislation on multiple grounds. In addition, the European Union has implemented stringent controls with regard to the privacy of member countries' citizens. Outside the United States and the European Union, the potential for using unique identifiers for disambiguation of individual humans appears to be problematic at the least and practically impossible at the worst. What to do in the absence of universal identifiers raises multiple kinds of questions for law enforcement, intelligence gathering, privacy advocates, and operational personnel charged with determining identity with a high degree of confidence.

2.5 INFORMATION OVERLOAD AND OTHER COMPLICATIONS OF HUMAN LANDSCAPES: THE CASE OF THE STASI

An extremely large-scale, holistic attempt to map the human landscape was undertaken in East Germany between 1950 and 1989. The lessons of the Stasi might inform current efforts insofar as it collapsed under its own weight and generated a wide variety of second-order consequences for which the ruling government was wholly unprepared. While many of these consequences related to the system of government that was in place, others grew out of the unsatisfactory solution to the fusion problems introduced by extensive surveillance.

2.5.1 Historical Context

The Ministry for State Security, commonly known as Stasi, was founded in East Germany in 1950 and performed both external and internal intelligence functions, gradually shifting resources toward the latter. By the 1980s, the number of civilian informants soared to unprecedented levels: 91,000 professional employees were supplemented by a reported 174,000 informants to monitor a population of 16 million people in 1989. A BBC report in 2007 quoted upgraded estimates to over 2 million: one informer for every seven citizens. Nine out of ten informers were male, over 25 and under 40 years of age. The pay was meager, so money did not appear to be a primary motivator [21].

2.5.2 Information Overload

With so many everyday citizens spying on family members, coworkers, and neighbors, the quality of observations varied widely. Tina Rosenberg, in her book *The Haunted Land,* vividly describes the often uselessly granular material reported [22]:

> The Stasi knew where Comrade Gisela kept the ironing board in her apartment . . . and how many times a week Comrade Armin took out his garbage and what color socks he wore with his sandals while doing it . . . The Stasi kept watch on trash dumps and lending libraries—the names of those who checked out books on hot air balloons or rock-climbing equipment were of particular interest—and tapped the booths of Catholic confessionals and the seats at the Dresden Opera. Stasi cameras monitored public toilets. . . . Some of its dossiers on East Germans had a hundred categories of information . . . .

Among the *Inoffizieller Mitarbeiter* (unofficial colleague, the term for citizen informers used after 1968) reports, 90% contained hearsay evidence. A reported B said something about C. Sources were routinely obscured or stripped at low levels, giving superiors little sense of a given report's reliability or consistency. Handwriting samples, logs and taps of telephone calls captured at the switch, photocopies of intercepted mail, and even a collection of smalls (to connect a known individual to a piece of evidence such as an antigovernment handbill) added to the massive body of records.

In the end, the Stasi could not use the masses of data to maintain power. In the words of one historian [17],

> Despite its intellectual property amounting to tens of thousands of kilometers of documentation and tapes and despite its elaborate checks on the accuracy of its data, the ministry was plagued by the cognitive problems typical of all intelligence agencies . . . The sheer mass of material not only threatened to overwhelm the ministry's operatives—inundation being the price to pay for very long ears—but also compounded the problems inherent in cognitive analysis and operationalization.

2.5.3 Consequences of Massive Surveillance

After the fall of the Berlin Wall, East Germans began to grasp the depth and breadth of the *Inoffizieller Mitarbeiter* program and overran the Stasi offices as part of the Peaceful Revolution of 1989. While a few records were shredded, the vast majority were saved. After debates over national security (of states other than East Germany) and the intensity of the revelations, the records were made available for public inspection. Later, some of the shredded records were reconstructed, and a ministry placed in charge of their cataloging and access. In the aftermath of both broad public questioning (of the Stasi's role in Olympic steroid management) and countless expressions of private pain at being betrayed by a spouse or colleague, the path forward for the East German people has been difficult.

2.6 INFORMATION FUSION AND THE HUMAN LANDSCAPE

Involving as it does both static and dynamic, tangible and intangible, and human and nonhuman elements, the human landscape fusion problem introduces both new scales of existing techniques and the need for entirely new techniques. Both the math and the theory grow in complexity beyond the current state of the art, stressing fusion systems at every stage.

2.6.1 Fusion of Existing Data Sources

The problem of information fusion for characterizing the human landscape is challenging in its own right. While characterization and fusion of information related to the physical landscape is difficult, the techniques are well known and extensively researched [23]. In the problem of fusing data for the physical landscape, we are faced with a situation in which primarily physical sensors observe physical attributes of the environment (e.g., spectral radiance, acoustic emissions), and these scalar or vector quantities can be related to a state vector that represents the physical environment (through an observation equation). Ordinarily we cannot solve this problem directly, transforming from observations to a state vector, but we can usually solve the inverse problem. That is, given an estimate of a state vector that characterizes an observed entity, we can usually predict what would be observed via an observation equation subject to observational error. The fusion problem for the physical landscape becomes one of collecting a set of redundant, noisy observations (namely, more observations than are strictly needed to obtain a state vector), and estimating the value of a state vector that best represents the observations. The process for this procedure is summarized in Chapter 1, and described in detail by texts such as Hall and McMullen [24]. Thus, while challenging, the traditional fusion problem involves well-known mathematical techniques. It should be noted that common target tracking estimation problems may entail estimating a state vector involving three to nine components such as three components to specify a target's instantaneous position and three additional components to specify the instantaneous velocity. In related applications, however, the state vector may become very large with several hundred components. For example, the NASA space tracking estimation may seek to estimate the position of a satellite or spacecraft, and add factors to estimate coefficients of atmospheric drag and to estimate the coefficients of the Earth's geopotential. Clearly the ability to estimate such large state vectors depends upon factors such as the amount and type of data, interrelationships among the state vector components, and other factors.

By contrast, the fusion of information for the human landscape begins with issues such as: What does it mean to define a "state vector" to characterize the human landscape? How do/can we link observables into this state vector? (What is the equivalent of the observation equation(s)?) How does the state vector change as a function of time? (What is the equivalent of an equation of motion?) What types of representation schemes and mathematics are useful? A commonly emerging representation technique is to focus on network representations (e.g., to represent individuals and links between individuals). Then, subnetwork matching techniques and network dynamic tools can be used. However, this is by no means an all-inclusive representation or estimation method, despite its mathematical elegance. Extensive work must be done to define quantities such as "observables," "state vectors," "observation equations," and "equations of motion" concepts that apply to the human landscape.

2.6.2 Fusion of Emerging Data Sources

The task of information fusion for data related to the human landscape begins with the difficult matter of understanding what can and cannot be collected in a given scenario. Asking, "what are people thinking about the stability of the current government?" is a very different matter in France than in Somalia. Determining the required resources and, indeed, the very feasibility of obtaining human landscape information is an entirely different problem from assessing a physical landscape and its ability to be sensed. After getting some sense of what can be known or even reasonably guessed at, the needs for the mission at hand can be assessed and matched with potential information resources.

Managing the collection of human landscape information is yet another hurdle to be cleared. The need for nuance, the sensitivity of source identification, the complexity of triangulation, and the difficulty of disambiguating human individuals or groups (as opposed to physical locations with GPS coordinates, for example) all contribute to a tremendously difficult phase in the fusion process. Centralizing, federating, and/or distributing the collected data presents a further challenge as each type of information architecture introduces strengths and weaknesses into both the collection process and subsequent activities.

Taking a vast amount of extremely noisy data of variable reliability and compressing it into a manageable stream, moving from bits to semantics, has never before been achieved at any scale. Yet again, information fusion within a human landscape will have to both rewrite the rule book and reinvent the technology landscape to become feasible.

As if collection and compression were not sufficiently daunting, getting meaning out of any system of human landscape fusion will require these systems to perform reasoning in two dimensions simultaneously. That is, both top-down, hypothesis-driven and bottom-up, data-driven reasoning will need to be supported. Prior experience in the commercial business intelligence (BI) market suggests that working in both directions will introduce a new generation of management challenges. Rules engines can easily be tuned to identify credit risks or patients obtaining too many painkiller prescriptions, for example, but the state of both the banking and health care industries proves that effective operational procedures do not necessarily drive effective strategy.

Finally, the fused information relative to the situation(s) must be connected to decision-makers responsible for those various situational operations. The degree to which human conditions, sometimes assessed by either individual or human observers/analysts, can be conveyed in regularized ways to decision-makers at various levels of risk, abstraction from conditions on the ground, and known reliability, remains as a significant task for the decades ahead.

2.7 CONCLUSION

Observing, characterizing, and predicting the human landscape are a very challenging problem. The remainder of this book addresses issues such as data visualization and pattern recognition, methods for intelligence preparation of the human landscape, analysis, and collaboration methods, as well as understanding the nature of human observations that provide input data. The role of understanding the human landscape is becoming increasingly important from military applications to disaster relief, to monitoring environmental and worldwide health conditions. Fundamentally, it is necessary to understand and model the human landscape to obtain predictive capability for understanding both the evolution of human-induced events (e.g., IEDs to terrorist events) and the evolution of crises such as response to natural disasters.

References

[1] Rheingold, H., *Smart Mobs: The Next Social Revolution,* New York: Basic Books, 2004.

[2] Kipp, J., et al., "The Human Terrain System: A CORDS for the 21st Century." *Military Review*, September–October 2006, pp. 8–15.

[3] *U.S. Military Human Terrain Team Handbook,* September 2008, http://wikileaks.org/wiki/US_military:_Human_Terrain_Team_Handbook_Sep_2008.

[4] Benkler, Y., *The Wealth of Networks: How Social Production Transforms Markets and Freedom,* New Haven, CT: Yale University Press, 2006.

[5] Barabási, A. L., *Linked: How Everything Is Connected to Everything Else and What It Means,* New York: Plume, 2003.

[6] Tapscott, D., *Grown Up Digital: How the Net Generation Is Changing the World.* New York: McGraw-Hill, 2009.

[7] Brainard, L., and D. Chollet, (eds.), *Global Development 2.0: Can Philanthropists, the Public, and the Poor Make Poverty History?* Washington, D.C.: Brookings Institution Press, 2008.

[8] Bick, J., "*Write a Check? The New Philanthropist Goes Further,*" *New York Times*, March 18, 2007.

[9] Armendáriz, A., and J. Morduch, *The Economics of Microfinance,* Cambridge, MA: MIT Press, 2005.

[10] Shirky, C., *Here Comes Everybody: The Power of Organizing Without Organizations,* New York: Penguin 2008.

[11] Kahn, J., "Mumbai Terrorists Relied on New Technology for Attacks," *New York Times*, December 9, 2008.

[12] Holland, J., *Hidden Order: How Adaptation Builds Complexity*, New York: Basic Books, 1996.

[13] Farid, H., "Digital Image Forensics," *Scientific American*, June 2008, pp. 66–71.

[14] http://www.biz.uiowa.edu/iem/research/LongRunAccuracy/.

[15] Kerner, N., and G. Pressman, *Chasing Cool: Standing Out in Today's Cluttered Marketplace*, New York: Atria, 2007.

[16] Engel, C., "El Al's Security a Model for Airlines," *Chicago Tribune,* September 10, 2006.

[17] Polanyi, M., *Tacit Dimension,* Chicago, IL: University of Chicago Press, 2009.

[18] Shapiro, L., *Perfection Salad: Women and Cooking at the Turn of the Century,* New York: Modern Library, 2001.

[19] Jones, J. H., and A. C. Kinsey, *A Life,* New York: W. W. Norton, 2004.

[20] Ferguson, N., *The Pity of War: Explaining World War I,* New York: Basic Books, 2000.

[21] Schmeidel, J. C., *Stasi: Sword and Shield of the Party,* New York: Routledge, 2008.

[22] Rosenberg, T., The Haunted Land, *Facing Europe's Ghosts After Communism*, New York: Random House, 1995, pp. 290–291.

[23] Liggins, M., D. Hall, and J. Llinas (eds.), *Handbook of Multisensor Data Fusion, 2nd ed.* Boca Raton, FL: CRC Press, 2008.

[24] Hall, D., and S. McMullen, *Mathematical Techniques for Multisensor Data Fusion, 2nd ed.* Norwood, MA: Artech House, 2004.

Chapter 3

H-Space: Humans as Observers

David Hall, Nathan McNeese, and James Llinas

The worldwide proliferation of cell phones and the evolution of cyberconcepts such as Facebook, Twitter, and News To Me enable humans to act as observers (soft sensors) for information fusion systems for improved situational awareness. But a paradox exists. On one hand, humans are clearly inferior to traditional hard sensors, such as radar, in observing physical objects like moving targets and estimating their position, velocity, and attributes. On the other hand, humans have a unique ability to observe and characterize other humans: inferring intent, motivation, and other quantities cannot be observed using traditional sensors. Hence, despite their inferiority with respect to observing physical targets, humans can provide significant value in observing the human landscape, to complement the use of hard sensors for observing the physical landscape. However, in order to fuse these human inputs with traditional (hard) sensors, it is necessary to understand the performance of human observers. This chapter introduces the problem of characterizing soft sensors, develops a framework for modeling performance, and provides a review of the literature on human observational performance.

3.1 INTRODUCTION

Rapid changes are occurring in how information is acquired and reported for events ranging from natural disasters to breaking news events. The worldwide proliferation of 4 billion cell phones (with embedded sensors including video and image capture, Global Positioning Satellite (GPS) geo-positioning, and other sensor devices), new collaboration environments such as Twitter, micro-blogs, News To Me, Facebook, and Flickr enables every human to become a potential observer/reporter. Individuals can act as both sensor platforms (acting as a mobile unit that transports a cell phone multisensor system) and observers who report on events via text messages input to a cell phone or mobile computing device. Ad hoc groups of human observers can augment traditional hard sensors used in information fusion systems. Hall et al. [1] introduced the concept of human-centered information fusion systems, in which humans act in three new roles: (1) as ad hoc observers and primary sources of information to augment traditional hard sensors, (2) as cognitive partners in the automated computer-based reasoning process to support pattern recognition and semantic reasoning for improved situational awareness, and (3) as ad hoc collaborators to perform dynamic problem analysis using social networks and virtual world technologies [2].

The effective use of human input in fusion systems requires the characterization of human observer performance and transformation of semantic data reports into quantitative values for state estimation. Failure to accurately characterize the performance of information sources and the subsequent degradation of fused results is one of the "dirty secrets" of data fusion cited by Hall and Steinberg [3]. This chapter describes issues in characterizing the human observer. Other researchers such as Antony [4] and Auger and Roy [5] have discussed how to translate fuzzy linguistic reports into quantitative scalar values and associated uncertainties.

It should be noted that currently there is no universally accepted definition of the term soft sensors. The origins of the use of the term "soft sensor" as used in this book to represent human observers can be traced to several sources, including the 7th Workshop on Information Fusion held at Beaver Hollow, New York, hosted by the Center for Multisource Information Fusion [6], and special sessions on Fusion Human Observations and Traditional Sensor Data held at the 11th and 12th International Conferences on Information Fusion (available from [7, 8]). The attendees at the Beaver Hollow workshop [6] debated about several possible definitions, but did not reach a consensus. We use the term here to indicate that humans generally act to provide an interpretation of their physical sensors to result in an observation or declaration (e.g., "I see a tall man near the bank"). The declaration is often stated in terms of a fuzzy semantic term (e.g., "near" or "tall") and has associated uncertainty which is difficult to quantify, since the observation often involves both a translation of physical senses and a cognitive assessment.

3.2 SOURCE CHARACTERIZATION CHALLENGES

It could be easily argued that humans are relatively poor observers. Experiments involving the use of distracters demonstrate that human observers can be surprisingly ineffective observers if they are distracted or are not anticipating particular events. In an experiment conducted by D. Simmons and C. F. Chabris [9] at the University of Illinois, for example, subjects were shown a video clip of two teams of people (in white and black uniforms) passing a basketball back and forth. The subjects were asked to count the number of passes by one team during a 60 second period. During the video, a person in a gorilla suit walks between the players, stops and waves, and continues on. A surprisingly high number of subjects fail to notice the gorilla (http://viscog.beckman.illinois.edu/flashmovie/15.php).

Other experiments conducted by Simmons and Rensink [10] involve a person change. In this class of experiments, an experimenter approached a pedestrian to ask for directions. While being provided directions, two additional experimenters rudely passed between the initial experimenter and the subject pedestrian. During this interruption, the initial experimenter was replaced by a different person, who looked quite different (and had a distinctly different voice) from the original experimenter. After the switch, approximately 50% of the subjects failed to notice that they were talking to a different person!

Despite these experiments, there are counterexamples in which humans are remarkably capable observers, especially in characterizing the human landscape. Indeed, our brains appear to be hardwired to recognize and characterize human concepts. Kaplan and Kaplan [11, p. 81] point out that "we can infer animal—specifically human-motion from remarkably few clues: a shifting pattern of white dots on a black background. When these moving dots mark the position of joints of a walking person, the observer spots it immediately—and can even specify the size, sex and, indeed, mood of the person with surprising accuracy." Yet Kaplan and Kaplan also point out humans can be readily deceived and predisposed to see unlikely things in unlikely places. They cite the following example [11, p. 82]: "In 1978 the Rotterdam Zoo reported the escape of one of its red pandas; hundreds of helpful people called in, having spotted it in places all over the Netherlands—when in fact it had been run over by a train just a few yards from the zoo fence."

Human observations are affected by stress, fatigue, drugs, emotion, expectations, prior training, and social impact of other observers [12, 13]. Other factors include experience (e.g., recognition primed effects [14]) and familiarity. In addition, the ability to retain information and accurately report decays rapidly with time [15, 16]. Our memory is more akin to an imagining of what might have happened, rather than an accurate recall of events [15]. Despite these challenges, the use of human observers can be effective, particularly to understand the human landscape. An example was cited by police officials in Roseville, California, who enlisted the aid of sanitation workers (in a program called "waste watch") to assist in monitoring local neighborhoods to report any unusual behavior, events or activities. The sanitation workers acted as extended mobile observers for the police, providing improved situational awareness and detection of anomalies (http://www.nytimes.com/2009/01/18/us/18trash.html). Other examples are provided by Gladwell [17].

Llinas [18] has pointed out that humans are the only means of providing information related to human landscape issues such as intent. Examples of the use of hard versus soft sensing are relevant for characterizing both the physical landscape as well as the human landscape. An example of characterizing a physical entity (for instance, a building) would include obtaining information, such as *building name*, *alias, address*, geolocation (latitude and longitude), boundaries, extent (e.g., people capacity, area), floor plan, *purpose*, *schedule of use, ownership* (who owns the building), and so forth. The attributes shown in italic text are most easily obtained by human observers, while the other attributes may be more readily obtained by physical sensors. Similarly, information about an individual human may include information, such as name, DNA, height, weight, hair color, fingerprint pattern, *current attitude and mood*, gender, *attractiveness, honesty*, and so forth. Again, some of these attributes can be more easily obtained by human observers rather than physical sensors. Thus, situational awareness of both the physical and human landscapes may require both physical sensors and human observers.

3.3 A FRAMEWORK FOR CHARACTERIZING OBSERVERS

Human beings are complex assemblages of physical and cognitive capabilities, emotions, and interactions with other humans. A framework for human-centered fusion must take into account this complexity before structuring appropriate modes of tasking, collection, fusion, and decision support.

3.3.1 A Conceptual Framework

A framework for understanding human observations is illustrated in Figure 3.1. A human observer is analogous to a conventional sensor that transforms energy (e.g., infrared, acoustic, electromagnetic energy) into an observational vector to provide information about target location, velocity, attributes, or derived information, such as identity, based on observed features. Similarly, human observers transform

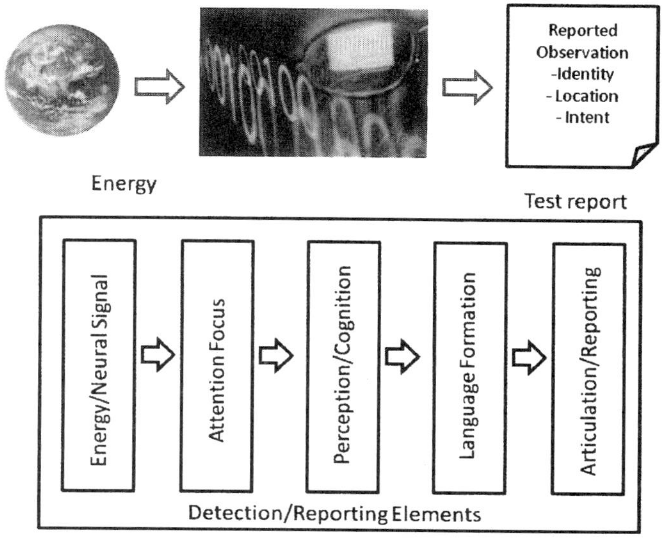

Figure 3.1 A conceptual framework for understanding human observations.

energy impinging upon their senses into verbal or textual reports. The output could be expressed as an aural output (spoken or musical expression), a graphic (diagram, pictograph, and icon), or even a gesture. The model identifies five key processes: (1) transformation of received energy into a neural signal in the brain, (2) attention paid to the received signal, (3) perception/cognition to translate the observation into a cognitive entity (thought or feeling), (4) transformation of the cognitive entity into language, and (5) articulation of the language construct into a text message or verbal utterance. These steps are an artificial partitioning of a dynamic, continuous process. However, they are useful for discussing how the transformation may occur and understand how to characterize an observer. For example, while there is little we can change regarding the transformation from light or sound into neural impulses (namely, the eye to neural impulse transformation or sound to neural transformation), we can understand and model some of the limits of such a transformation [19, 20].

Before discussing this model in detail, it should be acknowledged that this model is strictly a conceptual framework rather than a predictive model of human sensing and reporting or even an accurate model of the sensing and cognition process. Clearly, the human sensing, cognition, and reporting process is very complex. Our senses operate in coordination with each other and provide us with an awareness of evolving situations and threats without overwhelming us with sensory impressions. Kaplan and Kaplan [11] point out the cocktail party effect, where "we can follow a conversation through a welter of similar noise that would defeat any microphone—amazingly, the brain adjusts perception so that speedy light and slow sound appear to arrive in sync for things up to thirty meters away" [11, p. 83]. Ramachandran [21] describes experiments that clearly demonstrate that what we perceive as "now" has actually occurred 0.25 to 0.5 second prior to our perception. In addition, Ramachandran provides examples of neurological disorders such as "blind sight" in which a person apparently can accurately point to objects that they cannot "see." This, coupled with concepts such as synesthesia in which senses are "crossed" [22], indicate that our sensing/perception system is complex and not a simple energy for which to report process.

Thus, the model presented here is not a true biological and informational model of how sensing, perception, and reporting work, but it is intended to be useful to support an understanding of the issues related to utilizing and modeling human soft sensor reports.

3.3.2 Energy to Neural Activity: A Brief Survey of the Human Senses

As with physical sensors, ultimately, our awareness of the world is based on transforming energy (electromagnetic, acoustic, chemical, and so forth) into a signal, in our case, neural activity in the brain. We provide here a brief survey of the human senses.

Vision

As with all human senses, the process of human vision is complex. It extends beyond the optics of the eye to include the perceptual process after energy is transformed by the eye into neural activity. The act of seeing involves not only an optical process but also a perceptual process. We observe a three-dimensional world via a two-dimensional (upside-down) image on our retinas. Our binocular vision provides only a limited ability to determine distance. Thus, our perception of the three dimensional, complex world relies on processing that appears to involve at least partially inborn and partially inherent "naïve visual physics" [23]. For example, the peekaboo game universally played with human children seems to be designed to teach them that when object A is occluded by object B, object A still exists in the world, even though it is not still visible. Interestingly, computer game designers use such naïve visual physics to improve the apparent reality of their games—in a computer game, when object A occludes object B, there is no need to continue the computational expense of maintaining a representation of object B; it can simply disappear from the virtual world, and the game player will unconsciously assume its continued existence.

Studies of people who have their sight restored after a long period of blindness indicate that some aspects of vision are difficult. Fine et al. [24]

described a patient (identified as MM) who was blinded at 3½ years of age and had his sight restored at the age of 43. The patient had difficulty with three-dimensional interpretation of retinal images and difficulty in recognizing patterns, including face recognition. After sight restoration, he was able to recognize only about 25% of common patterns. By contrast, he was successful at many motion tasks, including determining the direction of simple and complex motion of objects. This suggests that some visual concepts are more inherent than others. It might be, for example, that identification of moving objects represents a fundamental characteristic to improve survival. It has been reported that certain types of martial artists learn to change their focus and even deliberately blur their vision to improve reaction time to potential attacks. Kaplan and Kaplan [11, p. 89] stated, "Adepts at kendo, Japanese fencing, practice a discipline called, 'looking at the mountain'; maintaining a wide, unprejudiced visual attention—since an incipient attack may signal itself as much in the twitch of a toe as the blink of an eye."

Other evidence of the role of perception in vision is provided by Oliver Sacks [25]. He described the case of an adult male painter who, due to an automobile accident, lost his ability to perceive color. While he intellectually knew he should perceive colors such as yellow or blue, he did not experience that perception after the accident. Interestingly, he could perceive the spectral differences that would be associated with different colors, but did not have the color perception. Sacks [25] provided examples of paintings done by the painter before and after his accident. His attempts to paint in color after the accident are especially interesting. Ultimately, the painter came to appreciate a new level of visual acuity that came with his color perceptual blindness.

Vision perception is apparently a highly parallel process. Kaplan and Kaplan [11] noted that while the brain has a relatively slow processing rate (neurons can only fire approximately every 5 milliseconds), only about 100 "calculations" can be performed in the half second notice-decide-act cycle. This implies that a high degree of parallel processing is required to deal with the 10 megabits of data that our vision processes per second. Kaplan and Kaplan cited research by Goodale and Milner [26], which supported two separate visual pathways in the brain; the ventral system processes pattern recognition to identify objects and patterns, while a separate dorsal system handles vision issues related to "where is this?" and "what can I do with it?"

The transformation of electromagnetic energy to neural activity involves light entering through the eye's lens and impinging upon the retina, which contains approximately 120 million rods for black and white vision and 6 to 7 million cones for color vision. The eye's optical system is by no means a perfect optical system. It has an aperture (pupil size) ranging from 3–4 mm to 5–9 mm depending upon lighting conditions, emotions, and the effect of drugs and a focal length of 17–22 mm. The eye is plagued by spherical aberration and other effects that distinguish the eye from a perfect camera. The wavelength sensitivity is in the range of approximately 360 nm to 720 nm, and the angular limit of resolution is 1×10^{-4} radiens (corresponding approximately to the diameter of a human hair at a distance of 1 meter). So-called normal vision is 20/20, measured using the Snellen chart found in most physicians' offices. This chart provides a hierarchy of different sized numbers, letters, and characters to quickly determine the state of someone's visual acuity. At 20/20 vision, a person with normal vision can recognize an optotype (letter on the Snellen chart) when it subtends 5 minutes of arc. Pilots and some athletes commonly have vision at a level of 20/10. For reference, on a clear day, a human with normal vision can detect an object 1 foot above a level plain at a distance of 1.3 miles. The determination of vision problems such as color blindness, myopia, astigmatism, and other vision problems requires the use of more sophisticated tests and measurements.

As an optical system, the eye also suffers from problems including a blind spot and variable angular fidelity across our field of view. The blind spot is a readily demonstrable gap in our vision (see http://www.blindspottest.com/ for a number of examples demonstrating the blind spot). In essence, there is a surprisingly large location of your retina where the optic nerve attaches and where you have a blind spot. Rather than actually seeing a blind or dark spot at that location in our field of vision, the brain "fills in" the blind area based on the ambient background. Hence, we "see" what isn't actually there (but the "filled in" area is so plausible, it is unnoticeable). In addition to the blind spot, our visual acuity is quite variable across our entire field of view.

Despite readily identifiable issues with human vision, it remains a remarkable system. While there are various types of optical illusions (see, for example, the 83 types of optical illusions at http://www.michaelbach.de/ot/), we are able to navigate in a complex three-dimensional world in a wide variety of illumination conditions, angles of observation, and different observing environments. Moreover, we can function in a very wide range of light conditions from bright sunlight to faint starlight (a range of approximately 10 million to 1), and it has been estimated that the vision is equivalent to a 576 Mpixel camera system (see http://www.clarkvision.com/imagedetail/eye-resolution.html). Attempts to make computers "see" (computer vision) show the complexity and sophistication of the human vision system.

Hearing

Another powerful sense available to humans is their sense of hearing. As with vision, the overall hearing process involves both a physical transformation of energy (acoustic vibrations) into neural impulses, as well as a perceptual process. The physical process translates acoustic vibrations received at our ears into neural electrical impulses. In the outer ear, the vibrations (in the range from 20 Hz to 20,000 Hz) are received at the ear aperture and transduced via the tympanic membrane (commonly called the eardrum). The middle ear contains an auditory canal, which contains a three-bone structure resembling the shape of a hammer, anvil, and stirrup, which amplify the sound analogous. The inner ear is filled with fluid and contains a sensory epithelium tissue studded with hair cells. The hair cells are mechanoreceptors that release a neurotransmitter when stimulated, translating mechanical vibrations into neural impulses. These are sent via the cranial nerve to the cerebral cortex in the brain.

As with sight, hearing is significantly affected by ambient conditions. Because of atmospheric heating, sound is conducted vertically during the daytime (especially near noontime) and, as a result, the distance at which we can hear a specific sound (such as a human voice) is greatly diminished compared to the evening or nighttime. During the evening and night, sound waves are guided by the atmosphere along the ground and travel a much longer range than during the day. This can cause an effective change in hearable distance for a specific sound by a factor of 100 or more.

Our hearing limits [27] include a frequency range of approximately 20 Hz to 20,000 Hz, a frequency resolution of approximately 0.36 Hz in the octave range from 1,000 to 2,000 Hz, and an intensity range from 0 dB to greater than 120 dB. Here, the measure of intensity is the logarithmic decibel (dB) scale in which 1 dB corresponds to an intensity of 1.97×10^{-10} atmospheres. This represents a huge dynamic range, from the equivalent to hearing a jet engine at 25 meters distance (120 dB) to normal conversation levels equivalent to 60 dB, to very quiet breathing at a level of 10 dB. Factors such as frequency variation, anticipation, and habituation strongly affect our perception of sound. In particular, habituation (continual exposure to a sound) can cause us to be unable to hear the sound after a period of exposure.

There are numerous types of hearing impairments caused by physical damage to the ear (e.g., ear canal obstructions, damage to the tympanic membrane (the eardrum), damage to the middle ear ossicles (bones), and other anomalies); noise-induced hearing loss; age-related declines, including reduction of sensitivity, loss of frequency range, and decreased dynamic range; medication-induced hearing loss; and physical trauma and impairments induced by disease. Noise-induced hearing loss may be particularly pervasive in members of the baby boomer generation who attended rock concerts and continually listened to very loud music, in addition to people who work on aircraft flight lines and operate loud equipment. The recent ubiquitous use of iPods and MP3 players may create similar problems. A particularly challenging hearing impairment is tinnitus, which involves the continual perception of sound without a corresponding external source (e.g., a continual ringing or buzzing in the ears). Restoration of at least partial hearing may be obtained via introduction of artifacts such as hearing aids or surgical procedures such as cochlear implants or surgical repair of the tympanic membrane or ossicles. However, resolution of tinnitus cannot be obtained via such techniques.

Early hearing loss or congenital hearing defects generally impair the ability of a child to learn language skills. Although, there is evidence that children who receive cochlear implants to improve their hearing abilities can overcome language deficits and learn language. Interestingly, there is some evidence that the brain can partially compensate for hearing loss by allowing the subreading of the auditory-associated cortex to shift functionality from being directed towards interpretation of aural signals to language, to a visual process that links sign language gestures to language. Thus, unlike vision, in which a sight-restored person may not be capable of ever achieving recognition of common patterns or shapes, hearing-restored people do seem capable of learning language and speech skills.

The role of perception in the hearing process can be illustrated by the concept of aural illusions, analogous to the concept of optical illusions. S. Wright coined the term *mondegreen* [28] to describe how she had misheard the reading from a line in the seventeenth-century ballad, *The Bonnie Earl O'Murray*. Instead of hearing the correct phrase, "And Lady Mondegreen," she heard "and laid him in the green." Another type of illusory hearing involves oronyms, sentences that can be read in two ways with the same sounds. For example, the sentence, "the sons raise meat" has the identical sounds to "the sun's rays meet." Clearly, contextual interpretation is needed to understand and process the phrases.

Hearing is a powerful and sometimes undervalued sense (by those who have normal hearing). Helen Keller is reported to have said, "Blindness cuts you off from things, deafness cuts you off from people."

Smell

One might be tempted to think of the sense of smell as a very limited sense, primarily focused on evaluating foods for edibility (although that appears to be a culturally dependent and learned response). But smell can play a powerful role in our assessment of the external world, including the human landscape. Kaplan and Kaplan [11, p. 109–110] reported that Oliver Sacks in "The Dog Beneath the Skin," a chapter in [25], tells the story of a student who, after too many amphetamines, briefly enjoyed the same vivid and relentless quality of life—this time through the sense of smell. He noted the "happy" smell of water, the "brave" smell of stone; he could name who was in a room before he entered it. It was a huge pleasure but at the same time "a world overwhelming in immediacy where thought became difficult and unreal."

The sense of smell strongly affects human mate selection. Kaplan and Kaplan [11] related the results of a Swiss T-shirt study: "This was neatly confirmed in the famous Swiss T-shirt test of 1995; a variety of men were asked to sleep in the same T-shirt for three consecutive nights. A range of women were then given the shirts to smell and asked, on that evidence alone, which of the men they found attractive. The result was remarkable: each woman chose the man whose immune system (as measured by his major histocompatibility complexes) was most different from her own [29]."

The sense of smell involves sensory neurons that match airborne molecules of certain chemicals to preexisting detectors or odor detectors. There are multiple competing theories about how this odor coding and perception process actually works, including a *shape theory* that supposes that receptors detect features of odor molecules, a *weak-shape theory* (odotope theory) that proposes that different receptors detect only small pieces of the odor molecule and fuse these together into an odor perception, and a *vibration theory* that suggests that odor receptors detect frequencies of odor molecules in the infrared range by electron tunneling. It is interesting to note that humans have about 10 square centimeters of surface area in the olfactory epithelium compared to nearly 170 square centimeters for dogs. In addition, dogs' olfactory epithelium has about 100 times more receptors per square centimeters. Penn State researchers studying the fluid mechanics of dog noses (and trying to develop electronic nose equivalents) note that the "refresh" rate for smelling is about 5 Hz. Thus, dogs sniff at a 5-Hz rate to refresh the flow of odor molecules across their sensing mechanism (http://www.rps.psu.edu/ indepth/ onthescent.html). Even though human smell perception is very limited compared to dogs, the nose can recognize over 10,000 scents.

The loss of smell can have serious effects on mood and even survival ability. The online site Wrong Diagnosis (http://www.wrongdiagnosis.com/sym/loss-of -smell.htm) lists 67 causes of loss of smell including diseases and side effects. Categories of causes include physical changes to the nose, toxic damage to the nose (e.g., from smoking or chemical exposure), damage to the brain, or nervous system, and certain diseases. Loss of smell can lead to loss of appetite, loss of taste, inability to detect dangers such as chemical or gas leaks, spoiled foods, and other problems.

Touch

The sense of touch is an often ignored (or taken for granted) sense. The sense of touch originates in the skin via nerve endings specialized for touch and pressure (to detect movements or pressure), cold and heat receptors to sense temperature, and pain receptors that respond to a variety of stimuli. Our sense of touch is most sensitive in our hands, lips, face, neck, tongue, fingertips, and feet, with approximately 100 touch receptors in each fingertip. By contrast, we are least sensitive in the middle of the back. The importance of touch is easily underestimated. Robles-de-la-torre [30] argued for the need to incorporate a sense of touch in artificial environments and described the case of two patients (Ms. "G. L." and Mr. Ian Waterman) reported in the literature who had both experienced a permanent loss of touch throughout their entire bodies. The resulting difficulties in interacting with the environment were very challenging. Even though both had full vision, they experienced a very difficult time in interacting with the environment: from manipulating objects to experiencing food to controlling the orientation and motion of their bodies. Interestingly, the sense of touch is susceptible to illusions analogous to optical illusions. A special issue of *New Scientist* magazine [31] addressed the sense of touch and illusions such as Aristotle's illusion, change numbness, motion aftereffects, and perceptual rivalry. Emerging technologies are providing the ability to emulate a sense of pressure via haptic interfaces, and recent advances provide mechanisms for providing patients with artificial limbs a sense of touch. Reference [32] described the case of a 27-year-old woman with an artificial arm and hand, who was provided with a sense of touch by rerouting shoulder nerves to a device attached to her chest that replicated the sense of feeling in her artificial hand.

Taste

An overview of the sense of taste is provided by http://users.rcn.com/jkimball.ma.ultranet/BiologyPages/T/Taste.html. The sense of smell involves the ability to respond to dissolved molecules and ions that reach taste receptor cells, clustered in taste buds. Each taste bud has a pore that opens to the tongue's surface, allowing molecules and ions to reach the receptor cells inside. There are five main taste sensations: salty, sour, sweet, bitter, and umami. The first four sensations identified are familiar to almost everyone. The latter sensation of umami is the response to glutamic acid like monosodium glutamate found in many processed foods and in many Asian dishes. There have also been claims for additional taste senses such as a special taste for fatty substances.

The perception of taste is affected by a substance's temperature, visual appearance, certain compounds known as taste modulators, and characteristics of the individual such as age, hormonal influences, genetic variations, and other issues. Some people are known as "supertasters," possessing a sense of taste that is significantly enhanced compared to normal people. Up to 25% of the population may fall in this category. By contrast, some people may experience a loss of taste for a variety of reasons ranging from health problems, such as nerve damage or degenerative diseases such as Parkinson's disease and Alzheimer's disease, the effect of some surgeries (e.g., third molar extraction and middle ear surgery), and head and neck radiation treatments for cancer. Impaired or lack of taste can affect our ability to enjoy food and detect potential threats (e.g., poisons) and may lead to depression.

Sensory Integration

Figure 3.2 shows examples of how our portfolio of human senses combine to provide the capability for observing and characterizing the human landscape. Clearly our senses work closely together, performing an ongoing fusion process. Some senses are linked closely together, such as taste and smell; impairment of one of these senses seriously impedes the complement sense. Other senses can act

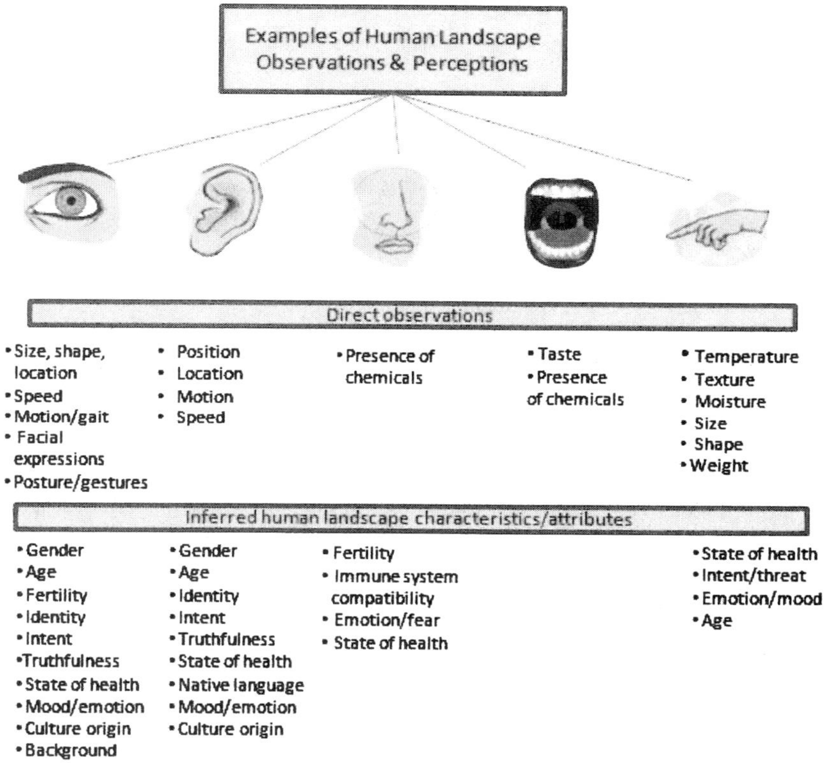

Figure 3.2 Example of human perception of human landscape.

as cues for others. For example, hearing can act as an early warning sensor to cue and focus vision, especially in limited seeing conditions. Finally, senses can work together in a gestalt mode to address issues such as detecting human intent or threats. Some of the inferences are intuitive, yet may be quite accurate, in part based on training. A trained physician may be able to make accurate inferences about the state of a person's general health based on combined touch, vision, and listening to the sound of a person's voice (regardless of the content of the speech). Cultural background may be inferred based on smell (e.g., some cultures consume foods high in garlic or glutamic acid), which can provide clues about culture even if a person cannot be seen or heard. While concepts such as "truthfulness" or "mood" may be difficult to ascertain, some people (sometimes referred to as "truth wizards") have remarkable skill at inferring these attributes.

The study of how humans integrate their sensory perceptions together is still emerging [33]. The concept of sensory integration was originally proposed by A. Jean Ayres in the 1970s. Ayres was an occupational therapist who introduced the concept that in order to function normally children need to learn to integrate information coming from multiple senses. In normal development, learning is believed to be "dependent on the ability to take in and process sensation from movement and environment and use it to plan and organize behavior" [33, p. 5]. Ayres' concept of sensory integration hypothesized that if one can introduce children to multiple sensory activities at once, the brain will attempt to integrate the feedback, correcting potential developmental problems. The developmental problems (called sensory integration dysfunction [DSI]) associated with the failure to develop effective sensory integration include: poor eye-hand coordination, auditory-language dysfunction, tactile defensiveness (an aversion to being touched), posture problems, frequent agitation, inability to focus attention, and other issues. DSI may contribute to Asperger's syndrome, autism, Tourett's syndrome, and other disorders. Ayres developed therapies designed to retrain children's neurobiological connections to improve the ability to smoothly process and effectively utilize stimulations from multiple senses. A challenge in understanding sensory integration is that the integration cannot be observed or self-reported directly (since it takes place at a nonverbal level). We can only observe the effects of disorders. Hence, evolving theory has been dependent upon research in neurobiology. Current research demonstrates that the sensory-cognitive integration is complex and is less of a hierarchy than once believed. Thus, progress in this area will need to ultimately seek an understanding of underlying subconscious neurobiological mechanisms, conscious experiences of sensory integration, and explanations of how childhood development and training result in normal versus dysfunctional development.

3.3.3 Attentional Focus

Attentional focus is the ability to direct one's attention cues in our environment that are relevant to a task at hand. While attentional focus is not completely understood, a failure to focus on a phenomena of interest is termed attentional blindness. We are "wired" to make rapid assessments of situations for survival purposes, but may ignore phenomena that would be readily apparent based on further attention and observation [34]. Interesting experiments described by Ramachandran [21] suggest that our intentions are subconscious and precede conscious intentional action by 0.25 to 1 second. When we consciously "decide" to conduct an action (such as observing something) the nerve impulses to accomplish the action have already been sent *prior* to our awareness of intention. This puts our consciousness in the role of a storyteller explaining what we have already apparently unconsciously chosen.

Assuming that we have received energy from the environment and "decided" to focus our attention on the observation, a perceptual/cognition process is required to provide a conscious awareness of the phenomena. This step may be bypassed if, for example, we receive a painful stimulus such as a touching a hot stove, quickly removing our hand before forming a cognitive experience of being burned. Another example is provided by the neurological syndrome called blindsight, involving patients with damage to their visual cortex. Such patients are reportedly able to point to, or identify, objects even when they do not consciously see them [21, p. 28]. Thus, there is receipt of energy, neural activity, and focus of attention, but no corresponding perception/cognition.

3.3.4 Perceptual Cognition

Human cognition is strongly affected by perception [34]. Perception and cognition are intertwined; perceptions based on sensory input are often changed to fit a pre-conceived mental model, and mental models affect what individuals are capable of perceiving. This interdependency motivates the term *perceptual cognition*.

Perceptional Velocity

Researchers have extensively studied characteristics that influence the relationship between perceived velocity and actual physical velocity. Fundamentally, a number of factors modify the way humans perceive speed [35]. The most influential factors are the size of the object, the nature over which the background moves, the contrast of the stimulus, and the stimulus spatial frequency [36, 27]. When looking at the effect size has on velocity, research has shown that a smaller object will appear to be moving at a higher velocity than a larger one [36, 37]. The reasoning for this is based on how much space the object takes up in our field of view (FOV). A small object takes up a small amount of space in the FOV, therefore leaving a large amount of background imagery to be seen.

The nature of the background imagery also plays an integral part on how fast an object is perceived to be moving. If a background is textured, it will significantly alter the perceived velocity [38]. Backgrounds generally are textured by landmarks. Landmarks in the background aid in perceiving motion, and decrease motion detection thresholds [36, 39–44]. Decreased motion detection threshold means that a slowly moving object might appear to be stationary if the background is lacking landmarks, but if the background is changed to include many landmarks, then that same slowly moving object is perceived to be moving much faster than before. Landmarks in backgrounds are not always stationary; sometimes they can be in movement. Norman et al. [45] demonstrated that when the background landmarks move only gradually faster than the object of focus, it is perceived that the object is moving at a much slower velocity than the background.

One aspect that also changes backgrounds and perception is contrast patterns. Suppose, for example, a ball is shown moving along a screen with dots on it, and the background also has dots in it. On a separate screen there is a ball with dots on it, but the background is made up of straight lines. Both of these balls are moving at the same speed, but we perceive that the ball with the dots and the lined background is moving faster. This is the Thompson effect—low contrast gratings are perceived as moving more slowly than high contrast gratings scrolling at identical speeds [46, 47]. A similar occurrence can be described using colors. A white ball will appear to move faster on a black background than a dark brown ball on a black background.

Spatial frequency of stimuli also affects perception of velocity. Increases in spatial frequency in low spatial frequency settings lead to an increase in perceived speed, while increases in spatial frequency in high-spatial frequency settings lead to a decrease in perceived speed [37, 48, 49].

Perceptual Distance

From infancy, humans learn to live in a world dominated by distance. The environment is composed of multiple objects that are in relation to each other based on this construct. Our perceptions of distance help us conduct everyday activities. Similar to perceptual velocity, there are varying factors that affect perception of distance. Elements that must be explained are the differences in interpretation of egocentric and exocentric distances, the differences between perceived distance in real and virtual worlds, and the power of the visual angle on perceptual distance.

Egocentric distance is the distance between the observer and a point in space. Exocentric distance is the distance between two external points. Studies testing these distances have traditionally shown that humans usually underestimate egocentric distance with an increase in the actual physical distance [50, 51]. Although exocentric distances are also underestimated, the likelihood of this happening is less. Experiments have demonstrated that perceived egocentric distance is much more accurate when *visually directed action* is used [51–54]. Visually directed action is a method used to give the observer subtle clues that help to understand the distance. These clues are usually in the form of visual cues. When visually directed action (cues) is decreased, errors become more abundant, meaning that the perceived egocentric distance is less precise. These results have also shown to be true when demonstrated on exocentric distances.

Recent research has addressed how perception occurs in virtual environments (VE). A main finding has been that humans tend to underestimate distance in VEs [55–57]. Most of this work has used head mounted displays (HMDs), which represent egocentric distance much more than exocentric. Although there have been significant findings that support the underestimation of distance, some researchers are wary of accepting this finding. They reason that

HMDs only provide a limited field of view in both horizontal and vertical directions. This limited view might be restrictive in providing the subject with a realistic view, therefore resulting in the HMDs' field of view being valued less.

Whether in a real or virtual environment, a key factor in the perception process is the visual angle used to observe the distance. Visual angles are the angle between the line of sight up to the mean point of a relative distance and the relative distance itself [58]. The relative distance can be changed depending on the angles from which the human is observing. If an observer is viewing a focal point at a sharp angle, a distance could be interpreted as being close. By comparison, if the observer views a focal point at only a slight angle, the distance will be perceived as being farther away. This is due in part because if the focal point is being viewed at an extreme angle, there are more contextual landmarks available to estimate distance than if the point is viewed straightforwardly or on a slight angle.

3.3.5 Perception of Language

While perceptions are important, they must be translated into language for communication to other humans. How is it that we see a car speeding down the road at an excessive speed, and are then able to say to someone, "That car is speeding"? Figuring out how we go from visual perceptions that are created in our brain to describing them through language is the problem that needs to be addressed.

A "language module" concept has been developed that many psychologists, including Steven Pinker [23], have chosen to advocate. This "module" is a structure located within our brain that allows the ability to store and process innate capacities for language. Within the language model, there are two schools of thought. First, the "module" is resistant to any incoming information that is not related to language [59]. Second, the "module" is part of a larger cognitive system that allows for broad collaboration between many regions within the brain [60]. A continuing debate involves whether or not the "module" is affected by other information coming into the brain. Currently, there has been more evidence to support the idea that there is collaboration between multiple parts of the brain. Through neurological scanning, it has been shown that almost any linguistic task activates many different brain regions [23]. These findings seem to refute the idea that the language module works by itself and is immune to other information not pertaining to language. While the idea of a language module is useful, it still does not address how we take visual perceptions and turn them into language. The best way to approach this is to look at the general factors influencing how we turn basic level perceptual recognition into higher orders of cognition.

The mind, as examined in a given moment, is certainly not a tabula rasa but comes with definite edges, biases, and connections that are: (1) driven by personal (or team-level) goals or intentions, (2) predicated by *human associative memory*

(3) significantly constrained by the context *being experienced,* and (4) strategically captured for use in the future (i.e., it can be learned and saved back to memory using specific kinds of *knowledge representations*). The way we work, think, and decide in the real world is dynamically synthesized and imprinted in the moment, but also informed by the past, and projected into the future based on intentions, expectations, and anticipations. A very important factor in translating perception to language is the role of experience. Basic experiences are translated into the mind on the grounds of what it finds to be similar through recognition once the sensory information has been stored. Once sensory information begins to be assimilated in the mind, basic experiences are recognized based on how similar they are to previously registered experiences. The best way to explain this is by looking at how an episode classifies our understanding of experience. An episode categorizes experience by recognizing experience based on cues or specific features that are familiar to us. By recognizing these familiarities, the mind allows us to experience the episode and stores it as an experience to later recall if need be. Another way that episodes are realized and activated, different from the use of specific features, is by language-based functions, such as declarative memories (encyclopedic, semantic-based associations, lexicons, and syntax rules produce language and thought in a more abstract even metaphorical level of comprehension). Along with declarative memories, there are also procedural memories that deal with temporal associations and motor-based actions. Both declarative and procedural associations provide the ability to link perception and language together.

Depending on whether a situation is a fast response or a slow response, certain language functions might be utilized or bypassed. In fast-response situations many of our innate language functions are bypassed because there is simply not enough time to go through the whole process. Even though we sometimes bypass many of the functions associated with language, we rarely are at a loss for words. This is because when a fast-response situation is activated we tend to significantly rely on our experiences of perception and language that we have stored from the past. Klein [61] showed that basic neural mechanisms allow the past to be associated with the present and stored for usage in the future. As would be predicted, the more experience one has with visual perceptions and language, the easier it is to formulate speech without actively knowing you are doing so. Experienced adults have encountered many visual settings and have created spoken language for these settings. The transformation of perception into language involves processes that span from practical sense making of direct perception (fast response) to symbolic metaphorical deep thinking of language (slower response). When encountering real-world situations, it is normal to try to make sense of the world through what we know (using both language and perception), but often our cognitive apparatus has to acquire new understanding. New understanding relies on an interpretive, meaningful analysis. This can involve deep cognition that uses higher-order processes such as articulation, social construction of knowledge (distributed cognition), creativity, and the conjoint nature of language and perception.

3.3.6 Language to Reporting

Once perception has occurred, language formation provides a means to articulate a thought to describe or characterize the observation. This, in turn, may be followed by articulation and reporting. Here again, these steps do not necessarily follow automatically. One can observe phenomena and develop perception through cognition, without necessarily being able to articulate the cognition in language. de Becker [62] provided examples of how fear can act as an alert system, providing valuable warning of threat conditions without language to articulate specific threats.

Verbal Overshadowing

Imagine that you are standing in line at the bank waiting to deposit your paycheck. Suddenly two men enter the bank, order everyone to get down on the floor, and demand that the bank tellers give them all the money. While lying on the floor, you get a direct look at one of the robbers' faces. You try to pay close attention to the details of his face and capture the image of it for later recall. Once the robbers have escaped, the police arrive and ask anyone if they saw the robbers. You gladly agree to help. You describe that the robber had a large forehead, high cheekbones, a small chin, and his hair and eye colors. A week later the police feel that they have some suspects that align with your description. The robber is within the lineup of suspects. The police now ask you to identify if any of the suspects was indeed the robber. To your amazement, you are now at a complete loss in identifying the robber.

This situation describes what has been identified as the verbal overshadowing effect. The act of describing a previously seen face will impair a later recognition of that face [63–65]. Schooler's initial experiment was set up so that subjects watched a bank robbery video. Half of the subjects were instructed to pay close attention to the robber and the other half was instructed to complete an activity that did not direct attention to the robber. The subjects who had been instructed to pay close attention to the robber were then told to describe the robber to the best of their abilities. Researchers found that the subjects who had been directing their attention on the robber and who had verbally described him were actually poorer at identifying the actual robber from a lineup than the subjects who had watched the video but completed a task not related to the robber. Schooler and Engstler-Schooler [64] described these results by arguing that the subjects who had verbally described the robber were then relying on their verbal representations more than their visual representations. The subjects who completed a separate task and did not focus on the robber used their visual representations, not verbal ones, and this was more successful.

This experiment suggests that verbal reports can interfere with the visual recall of an event, and that visual memory is stronger than verbal memory. When people use verbal representation to describe an event or a person, they are allowing their verbal bias to interfere with what was once an accurate visual representation. Verbal memory is usually very general, whereas visual memory tends to be more precise and detailed [66]. People using visual memory to identify something have many more facets that they can use to discriminate against certain characteristics as opposed to verbal memory. This implies that there may be unintended consequences of tasking people to immediately report their observations. Do we really want people verbally describing what they have seen? If verbal reporting is desired, are we willing to sacrifice information that could have been later reported, but the person cannot accurately recall because he or she already verbally reported something that causes interference? We must explore the best way to use humans so that the maximum amount of information is provided in the end result.

Source Monitoring

Source monitoring is concerned with how our experiences tend to shape our memories. When referring to source, it is the variety of characteristics that help to specify the conditions in which memory is acquired [67]. Source monitoring claims that there is not a label that specifies where a memory source is coming from, rather activated memory records are evaluated and then attributed to sources via a decision process that takes place during remembering [67]. For the purposes of using humans as soft sensors, this concept could be very important in understanding how to train people in remembering by emphasizing the use of sources to aid in memory. Source monitoring is something that is usually done automatically. When we start remembering, we seem to also identify the sources of the memory also. If you remember that you had a party for your tenth birthday, the cognitive processes automatically seem to start remembering things such as what the cake looked like, who was at the party, and what you got for your birthday. If these sources were not available for recall, then you most likely would not have any memory of your party.

The concept of source monitoring derived from the idea of reality monitoring. Reality monitoring refers to how we cognitively discriminate between external and internal sources of information [68]. Essentially what this means is how we differentiate perceived thoughts and imaginations (internal) from perceived events (external). This type of monitoring is looking at whether people have the ability, with some accuracy, to notice when a memory is imagined or is something that really happened. Research has been done on how accurate people are in differentiating these two types of memory and it has been found that older people tend to be less accurate than younger people when discriminating between real and imaginary [67].

Our previous experiences play a great role within how source monitoring works. The more experiences you have, the more activated memories your brain has to search through and evaluate to determine if that memory should be paired with a certain source. Much like reality monitoring, work has been done that looks at the correlation of source monitoring and age, and it shows that older people have a harder time determining sources because of the sheer amount of memories that they have collected over their lives [69]. Understanding why older people have a harder time with source and reality monitoring is not difficult. The more memories one has, the more difficult it is to differentiate the memory items, therefore making it harder to decide what the source is. This finding is interesting because we commonly believe that more experience automatically improves our perceptions and recall. Experience in source monitoring might not be the best because it confuses the definitions in our brain that help decide which source is the best for the memory. One could have 15 similar memories, but how does the brain differentiate those memories enough to define a source? All of these memories very well might end with the person attributing the wrong source.

Eyewitness Memory

Eyewitness memory could possibly be the most important factor that influences reporting. Its importance stems from the relation it has with almost every other factor. Without eyewitness memory there would be no verbal overshadowing effect, source monitoring, or memory decay. All of these concepts are defined in completely different ways, but they come back to one another full circle because they all start with eyewitness memory. When a person reports something, he or she is essentially relaying his or her eyewitness memory to someone else. Reporting in its purest form is eyewitness memory.

Because much research has been completed on this topic, it is a challenge to determine what to include and what to leave out. For the purpose of this chapter, the best approach is to describe what aspects in nature help us to remember some things and forget others. When a person is called to give an eyewitness testimony, it frequently revolves around the description of a person. How do we remember a person? What factors stand out more than others that we can recall at a later time? Examination of the research literature provides us with knowledge of how people remember others [70]. People seem to always remember primary descriptors such as height, weight, and age [71, 72]. This is not surprising because these are the most obvious characteristics of a person. One does not have to look hard to see that someone is tall and lanky in comparison to someone who is short and chubby. These are the type of characteristics that just pop out at us even when we are not proactively trying to evaluate a person's physical manner. Usually primary descriptors are reported with decent accuracy [73]. It is when people attempt to describe secondary descriptors that accuracy starts to decline. Secondary descriptors can be defined as hair color, eye color, and facial attributes such as beards. In eyewitness testimony, these are the descriptors that are the most important because they can help discriminate among multiple people. Unfortunately, people tend to not pay close attention to these descriptors [74].

People are not the only aspect that is important to eyewitness memory; events are as well. Events are remembered more effectively if they are perceived to be personally relevant and elicit physiological responses. A person is not likely to remember a car crash that happened a month ago if he or she was not directly affected by it. If the person was one of the drivers, he or she will most certainly remember the crash because it was very relevant personally and resulted in physiological responses.

Eyewitness memory for soft sensors is important, but it must be understood that there also can be a great deal of error involved. Even when we believe strongly that we experienced or witnessed something, it might not have actually happened [15]. Although we have the ability to imagine the past, for the most part, we are confident we can identify a person or event with relatively high accuracy if reported in the right time frame [16]. Humans are able to put meaning into what they have witnessed, but this might not always be for the best. Meaning can sometimes result in personal bias, which can lead to the poor reporting. Are humans the best way to report eyewitness memory? This question is open for discussion, but based on our sheer amount of experience, we certainly are the front-runners.

A general discussion of cognitive biases can be found in [75, 76]. Heuer [76] describes issues such as "seeing order in chaos," "looking for confirmation (confirmation bias) in perception," "reinterpretation of evidence (to fit preconceived notions)," and selective remembering of evidence. Again, while there are well-known biases in sensing and cognition, humans can still provide valuable evidence in assessing a situation or event—particularly because of their ability to account for contextual information.

3.4 SUMMARY

The framework presented here and the associated literature review suggest that a significant amount of work remains to develop meaningful models for the human observation process. Situational factors affect perception of distance, motion, and entity characteristics. Failure to properly focus attention can lead to situational blindness in which an observer simply fails to perceive significant activities or events. Despite the modeling challenges, the rapid adaption of cell phones and emerging collaboration tools such as Twitter suggest that human reports will be increasingly important for information fusion systems. As such, it is worthwhile to continue to develop models and frameworks for characterizing the soft sensor.

References

[1] Hall, D., et al., "A Framework for Hard/Soft Fusion," *Proceedings of the 11th International Conference on Information Fusion*, Cologne, Germany, June 30–July 3, 2008.

[2] Hall, D., et al., "Perspectives on Visualization and Virtual World Technologies for Multi-Sensor Data Fusion," *Proceedings of the 11th International Conference on Information Fusion*, Cologne, Germany, June 30–July 3, 2008.

[3] Hall, D., and A. Steinberg, "Dirty Secrets in Multisensor Data Fusion," chapter 21 in *Handbook of Multisensor Data Fusion*, D. L. Hall and J. Llinas, (eds.), Boca Raton, FL: CRC Press, 2001.

[4] Antony, R., "Data Management Support to Tactical Data Fusion," chapter 24 in *Handbook of Multisensor Data Fusion: Theory and Practice*, 2nd ed., M. Liggins, D. Hall, and J. Llinas, (eds.), Boca Raton, FL: CRC Press, 2008.

[5] Auger, A., and J. Roy, "Expression of Uncertainty in Linguistic Data," *Proceedings of the 11th International Conference on Information Fusion*, Cologne, Germany, June 30–July 3, 2008.

[6] "Hard and Soft Data Fusion, Fusion Frameworks and Enabling Technologies," *Proceedings of Workshop VII*, Beaver Hollow Workshop, New York, October 29–31, 2008, http://www.infofusion.buffalo.edu/.

[7] http://www.isif.org/.

[8] http://controls.papercept.net/conferences/conferences/FUSION09/program/FUSION09_ContentListWeb_3.html.

[9] Simons, D. J., and C. F. Chabris, "Gorillas in Our Midst: Sustained Inattentional Blindness for Dynamic Events," *Perception*, Vol. 28, 1999, pp. 1059–1074.

[10] Simons, D. J., and R. A. Rensink, "Induced Failures of Visual Awareness," *Journal of Vision*, Vol. 3, No. 1, 2003.

[11] Kaplan, M., and E. Kaplan, *Bozo Sapiens, Why to Err Is Human*, New York: Bloomsbury Press, 2009.

[12] Deffenbacher, K. A., et al., "A Meta-Analytic Review of the Effects of High Stress on Eyewitness Memory," *Law and Human Behavior*, Vol. 28, No. 6, 2004, pp. 687–706.

[13] Christianson, S., "Emotional Stress and Eyewitness Memory: A Critical Review," *Psychological Bulletin*, Vol. 112, 1992, pp. 284–309.

[14] Ross, K., et al., "The Recognition-Primed Decision Model," *Military Review,* July-August, 2004.

[15] Loftus, E. F., "Imagining the Past," *The Psychologist*, Vol. 14, No. 11, 2001, pp. 584–587.

[16] Odinot, G., and G. Wolters, "Repeated Recall, Retention Interval and the Accuracy-Confidence Relation in Eyewitness Memory," *Applied Cognitive Psychology*, Vol. 20, 2006, pp. 973–985.

[17] Gladwell, M., *Blink: The Power of Thinking Without Thinking*, New York: Little, Brown, 2005.

[18] Llinas, J., "Introductory Remarks on Hard and Soft Data Fusion," Hard and Soft Data Fusion Framework and Enabling Technologies, 7th annual Beaver Hollow Workshop, Beaver Hollow, NY, October 29, 2008.

[19] Moller, A. R., *Sensory Systems: Anatomy and Physiology*, New York: Elsevier Science, 2003.

[20] Ackerman, D., *A Natural History of the Senses*, New York: Vintage Books, 1990.

[21] Ramachandran, V. S., *A Brief Tour of Human Consciousness*, New York: PI Press, 2004.

[22] Cytowic, R., *The Man Who Tasted Shapes*, Cambridge, MA: MIT Press, 2000.

[23] Pinker, S., *How the Mind Works,* New York: Norton, 1997.

[24] Fine, A., et al., "The Effect of Long Term Deprivation on Visual Perception and Visual Cortex," *Nature Neuroscience*, Vol. 6, No. 9, September 2003.

[25] Sacks, O., *A Neurologists Notebook: An Anthropologist on Mars*, New York: Vintage Books, 1995.

[26] Goodale, M., and A. D. Milner, "Separate Visual Pathways for Perception and Action," *Trends in Neurosciences*, Vol. 15, No. 1, January 1992, pp. 20–25.

[27] Plack, C. J., *The Sense of Hearing*, Mahwah, NJ: Lawrence Erlbaum Assoc., 2005.

[28] Wright, S., "The Death of Lady Mondegreen," *Harpers*, November 1954.

[29] Wedekind, C., et al., "MHC-Dependent Preferences in Humans," *Proceedings of the Royal Society of London B.*, Vol. 260, No. 1359, 1995, pp. 245–249.

[30] Robles-de-la-torre, G., http://www.roblesdelatorre.com/gabriel/GR-IEEE-MM-2006.pdf, 2006.

[31] *New Scientist*, March 11, 2009.

[32] http://www.guardian.co.uk/science/2007/nov/27/medicalresearch/.

[33] Gelman, R., and T. Kit-Fong Au, (eds.), *Perceptual and Cognitive Development: Handbook of Perception and Cognition*, San Diego, CA: Academic Press, 1996.

[34] Hallinan, J. T., *Why We Make Mistakes: How We Look Without Seeing, Forget Things in Seconds, and Are All Pretty Sure We Are Way Above Average*, New York: Broadway Books, 2009.

[35] Brown, J. E., "The Visual Perception of Velocity," *Psychologische Forschung*, Vol. 278, No. 192, 1931.

[36] Darnell, L., "Illusions Disrupting the Accurate Perception of Velocity and Position," 2004, http://www.ucl.ac.uk/~ucbplrd/Lewis_Dartnell-MRes_project.pdf.

[37] Blakemore, M. R., and R. J. Snowden, "The Effect of Contrast Upon Perceived Speed: A General Phenomenon?" *Perception*, Vol. 28, No. 1, 1999, pp. 33–48.

[38] Snowden, R. J., "Perceived Speed: Effects of Field Size and Background Texture," *Investigative Ophthalmology & Visual Science*, Vol. 38, No. 4, 1997.

[39] Leibowitz, H. W., "Effect of Reference Lines on the Discrimination of Movement," *Journal of the Optical Society of America*, Vol. 45, 1955, pp. 829–830.

[40] Mates, B., "Effects of Reference Marks and Luminance on the Discrimination of Movement," *Journal of Psychology*, Vol. 733, 1969, pp. 209–221.

[41] Tyler, C. W., and J. Torres, "Frequency Response Characteristics for Sinusoidal Movement in the Fovea and Periphery," *Perception & Psychophysics*, Vol. 12, 1972, pp. 232–236.

[42] Legge, G. E., and F. W. Cambell, "Displacement Detection in Human Vision," *Vision Research*, Vol. 21, 1981, pp. 205–214.

[43] Johnson, C. A., and R. P. Scobey, "Effects of Reference Lines on Displacement Thresholds at Various Duration of Movement," *Vision Research,* Vol. 22, 1982, pp. 819–822.

[44] Bonnet, C., "Discrimination of Velocities and Mechanisms of Motion Perception," *Perception,* Vol. 13, 1984, pp. 275–282.

[45] Norman, H. F., et al., "Spatial Interactions in Perceived Speed," *Perception*, Vol. 25, No. 7, 1996, pp. 815–830.

[46] Thompson, P., *Velocity Aftereffects and the Perception of Movement*, University of Cambridge, Cambridge, U.K., 1976.

[47] Thompson, P., "Perceived Rate of Movement Depends upon Contrast," *Vision Research*, Vol. 22, 1982, pp. 377–380.

[48] Diener, H. C., et al., "The Spatial Frequency Effect on Perceived Velocity," *Vision Research*, Vol. 16, 1976, pp. 169–176.

[49] Smith, A. T., and G. K. Edgar, "The Influence of Spatial Frequency on Perceived Temporal Frequency and Speed," *Vision Research*, Vol. 30, 1990, pp. 1467–1474.

[50] Harway, N. I., "Judgement of Distance in Children and Adults," *Journal of Experimental Psychology*, Vol. 65, 1963, pp. 385–390.

[51] Loomis, J. M., et al., "Visual Space Perception and Visually Directed Action," *Journal of Experimental Psychology: Human Perception & Performance*, Vol. 18, 1992, pp. 906–921.

[52] Loomis, J. M., *Visual Perception of Egocentric Distance in Real and Virtual Environments*, Mahwah, NJ: Lawrence Erlbaum Assoc., 2003.

[53] Philbeck, J. W., and J. Loomis, "Comparison of Two Indicators of Perceived Egocentric Distance Under Full Cue and Reduced Cue Conditions," *Journal of Experimental Psychology: Human Perception & Performance*, Vol. 23, 1997, pp. 72–85.

[54] Rieser, J.J., et al., "Visual Perception and the Guidance of Locomotion with Our Vision to Previously Seen Targets," *Perception*, Vol. 19, 1990, pp. 675–689.

[55] Thompson, W. B., et al., "Does the Quality of Computer Graphics Matter When Judging Distance in Visually Immersive Environments?" *Presence: Teleoperators and Virtual Environments,* Vol. 13, 2004, pp. 560–571.

[56] Goch, A. A., and P. Willemsen, "Perceived Egocentric Distances in Real, Image-Based and Traditional Virtual Environments," *IEEE Virtual Reality Conference*, 2002.

[57] Plumert, J. M., J. K. Kearney, and K. Recker, "Distance Perception in Real and Virtual Environments," *ACM Transactions on Applied Perception*, Vol. 2, No. 3, 2005, pp. 216–233.

[58] Mathsushima, E. H., et al., "Visual Angle as Determinant Factor for Relative Distance Perception," *Perception*, Vol. 26, 2005, pp. 97–104.

[59] Foder, J., "So How Does the Mind Work," *Mind & Language*, Vol. 20, 2005, pp. 25–32.

[60] Rumelhart, D. E., and J. L. McClelland, "PDP Models and General Issues in Cognitive Science," in D. E. Rumelhart, J. L. McClelland, and the PDP Research Group, (eds.). *Parallel Distributed Processing: Explorations in the Microstructure of Cognition, Vol. I: Foundations*, Cambridge, MA: MIT Press, 1986.

[61] Klein, G., *Sources of Power: How People Make Decisions*, Cambridge, MA: MIT Press, 1999.

[62] De Becker, G., *The Gift of Fear: Survival Signals That Protect Us from Violence*, Boston, MA: Little, Brown, 1997.

[63] Fallshore, M., and J. W. Schooler, "The Verbal Vulnerability of Perceptual Expertise," *Journal of Experimental Psychology: Learning, Memory, & Cognition*, Vol. 21, 1995, pp. 1608–1623.

[64] Schooler, J. W., and T. Y. Engstler-Schooler, "Verbal Over Shadowing of Visual Memories: Some Things Are Better Left Unsaid," *Cognitive Psychology*, Vol. 22, 1990, pp. 36–71.

[65] Schooler, J. W., R. S. Ryan, and L. M. Reder, "The Costs and Benefits of Verbally Rehearsing Memory for Faces," in D. Herrmann, et al. (eds.), *Basic and Applied Memory: New Findings*, Hillsdale, NJ: Erlbaum, 1996, pp. 51–65.

[66] Dodson, C. S., M. K. Johnson, and J. W. Schooler, "The Verbal Overshadowing Effect: Why Descriptions Impair Face Recognition," *Memory & Cognition*, Vol. 25, No. 2, 1997, pp. 129–139.

[67] Johnson, M. K., S. Hashtroudi, and D. S. Lindsay, "Source Monitoring," *Psychological Bulletin* Vol. 114, No. 1, 1993, pp. 3–28.

[68] Johnson, M. K., and C. L. Raye, "Reality Monitoring," *Psychological Review*, Vol. 88, 1981, pp. 67–75.

[69] Henkel, L. A., M. K. Johnson, and D. M. De Leonardis, "Aging and Source Monitoring: Cognitive Processes and Neuropsychological Correlates," *Journal of Experimental Psychology*, Vol. 127, No. 3, 1998, pp. 251–268.

[70] Meissner, C. A., et al., "Person Descriptions as Eyewitness Evidence," in R. Lindsay, et al., (eds.), *Handbook of Eyewitness Psychology: Memory for People*, Mahwah, NJ: Lawrence Erlbaum & Associates, 2005, pp. 3–34.

[71] Kuehn, L. L., "Looking Down a Gun Barrel: Person Perception and Violent Crime," *Perceptual & Motor Skills*, Vol. 39, 1974, pp. 1159–1164.

[72] van Koppen, P., and S. Lochun, "Portraying Perpetrators: The Validity of Offender Descriptions by Witnesses," *Law and Human Behavior*, Vol. 21, 1997, pp. 661–685.

[73] Yarmey, A. D., and M. Yarmey, "Eyewitness Recall and Duration Estimates in Field Settings," *Journal of Applied Social Psychology*, Vol. 27, 1997, pp. 330–344.

[74] Ellis, H. D., "The Development of Face Processing Skills," *Proceedings of the Royal Society*, Series B, Vol. 335, 1992, pp. 105–111.

[75] Gilovich, T., *How We Know What Isn't So: The Fallibility of Human Reason in Everyday Life*, New York: The Free Press, 1991.

[76] Heuer, R., *The Psychology of Intelligence Analysis*, Washington, D.C.: CIA, 1999, https://www.cia.gov/library/center-for-the-study-of-intelligence/csi-publications/books-and-monographs/psychology-of-intelligence-analysis/index.html.

Chapter 4

Global Neighborhood Watch: The Emerging Community of Observers

The term "Global Neighborhood Watch" was introduced by Boisot and McKelvey in 2004 at an annual conference of the International Military Testing Association and subsequently published in the book, *Corporate Strategies Under International Terrorism and Adversity* in 2006 [1]. The idea extends the concept of citizens in local neighborhood watch groups seeking to reduce street crime to an international scale for reduction of terrorism. Similar concepts have been introduced to utilize urban sanitation workers as observers for police departments [2]. Citizen observers have also been called upon to support scientific investigations (e.g., monitoring bee populations [3] and earthquake effects [4]) and to solicit citizens to report newsworthy events to such organizations as the BBC and CNN [5]. The Web site Ushahidi [6] (Ushahidi is a Swahili term that means "testimony") allows users to generate reports overlaid on a Google Map to share information about epidemics, election results, and terrorist and crime activities. With the rapid growth in worldwide adoption of cellular telephony, and the use by sensor designers of the phone infrastructure as a platform, the disciplines of sensor science and practice are presented with new opportunities, constraints, and management issues. While these new rules are many, the most important of them revolve around the shift in constraints and the concomitant impact on management issues.

4.1 EMERGING TRENDS IN AD HOC GLOBAL OBSERVATIONS

The rapid rise on a global level of the cell phone as a widely available data capture, storage, and transmission device presents a variety of opportunities for information gathering.

4.1.1 Cell Phones as Sensors and Sensor Platforms

Depending on which estimate is used, the number of active mobile phones worldwide exceeded 3.3 billion in 2007 or 2008, ahead of earlier projections. While the devices are unequally distributed, the figure represents one subscription for every second person on the planet, and the industry is devising low-cost handsets and creative infrastructure solutions, such as biofuel-powered base stations, to continue to increase the technology's presence in the developing world [7]. Figure 4.1 shows the number of cell phones per 100 population for selected countries.

Interestingly, the United States lags behind a number of countries, including the Ukraine, Portugal, Russia, Germany, Estonia, Israel, and others.

At the same time that the number of handsets continues to increase (see Figure 4.2), researchers are deploying a wide variety of sensors on cell phones. In the process, the mobile phone is becoming a platform, which Michael Cusumano of MIT and Annabelle Gawer of Imperial College London have defined as "an evolving system made of interdependent pieces that each can be innovated upon" [8]. Some examples follow:

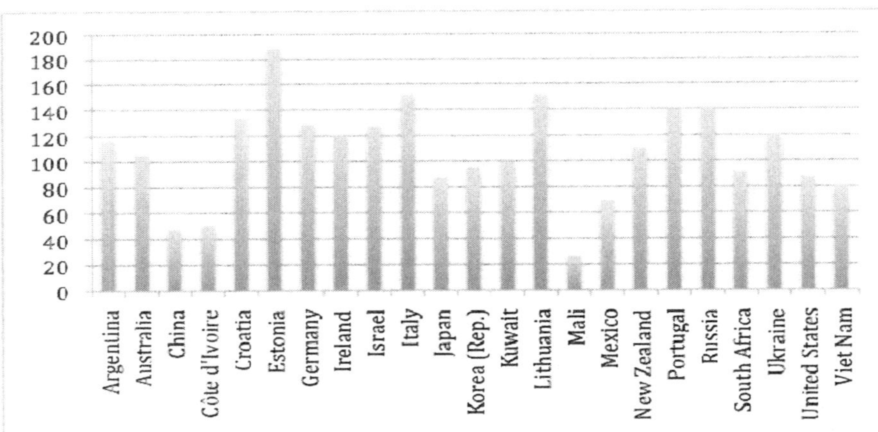

Figure 4.1 Mobile phones per 100 inhabitants, selected countries, 2008. (Source: ITU)

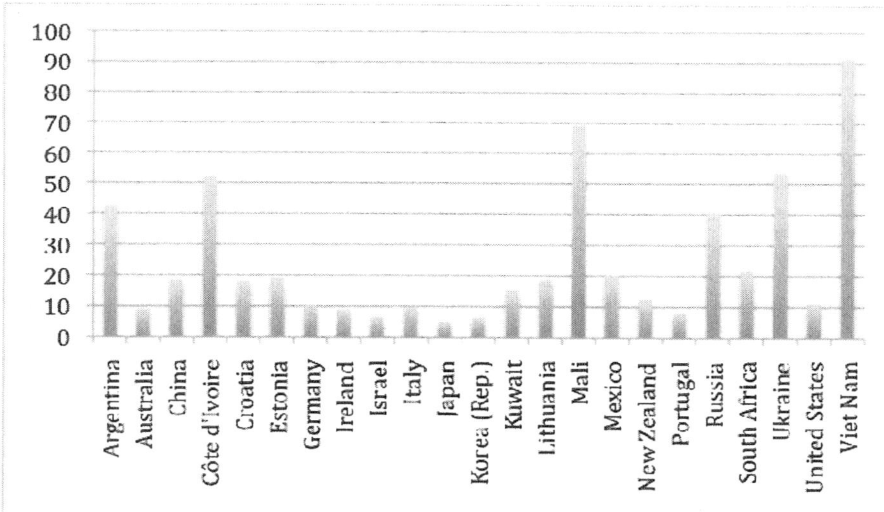

Figure 4.2 Cell phone teledensity growth rates 2003–2008, selected countries. Source: ITU.

- Ephraim Fischbach and Jere Jenkins of Purdue University are working to add low-cost, lightweight radiation sensors to mobile phones in order to detect the radiation characteristic of so-called "dirty bombs" [9].
- In February 2008, 100 vehicles carrying GPS-equipped Nokia phones were monitored to capture speed and location information for each vehicle at 3-second intervals. Compared to cameras, traffic helicopters, and other approximations, the technology promises more accurate information for both modeling and real-time applications [10].
- Nokia has built a prototype called the Eco Sensor Concept (Figure 4.3). A wearable unit, it detects environmental, health, and weather conditions, then wirelessly sends the data to a more powerful mobile phone that displays and, if appropriate, transmits the data. Potential uses include workout monitoring, ambient noise level detection, and environmental sensing of smog, toxins, and ultraviolet radiation [11].

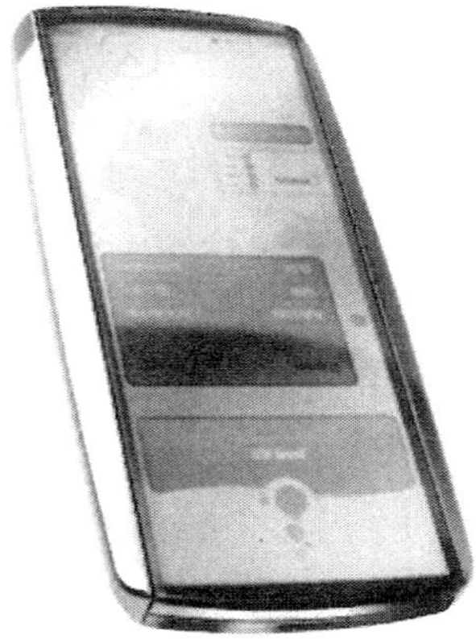

Figure 4.3 Nokia Eco Sensor Concept.

- The Blum Center for Disease Diagnosis at the University of California–Berkeley has built prototypes of diagnostic microscopes that transmit images of blood samples to health care personnel at a remote lab or other facility [12].
- Cell phone cameras were utilized by the BBC to provide rapid coverage of the July 7, 2005, subway, bombings in London [13]. CNN, Yahoo!, and Reuters later introduced similar programs to recruit citizens to submit photos and video [14].

4.1.2 Differences Between Cell Phones and Conventional Sensors

As a platform, the mobile phone enjoys several key advantages over traditional sensor networks. First, high transmission power and a large population of base stations mean that mesh technologies can be replaced with standard industry protocols. Second, compared to many sensor packages, the cell phone form factor allows for relatively large batteries, with easy recharging facilities. Finally, ample bandwidth allows for larger data payloads, without the need for dramatic economic routing techniques.

There are negatives to each of these attributes, of course. High transmission power, even if encrypted, facilitates easier detection. Larger form factors make intrusion in existing real-world processes more complicated. Competition for resources between the sensor bolt-ons and the phone's primary functions may not resolve in the sensors' favor.

Table 4.1 summarizes some of the differences between the two models.

Table 4.1
Comparison of Traditional Sensor Networks and Cell Phone Platforms

Traditional Sensor Network	*Cell-Phone Sensor Platform*
Unattended	Attended and/or worn on person
Time synchronization must be accommodated at cost in battery life, processor, and bandwidth	Time service available from platform; in the case of GPS-equipped devices, time is extremely accurate
Severely energy constrained, particularly wireless	Relatively large battery readily available
Limited power limits bandwidth	Robust bandwidth
Security may need to be built	Encryption included in base platform
1 × 1 × 1 cm form factor is common	Much greater latitude in packaging
Localization, if needed, must be generated on both relative (to other sensors) and absolute (latitude/longitude) levels	GPS and antenna-based localization already available

4.1.3 The Twitter Factor

A recent service that enables creation of dynamic observer networks using cell phones (or related devices) is Twitter. Twitter is a micro-blogging and social network service created in 2006 by Jack Dorsey that allows users to send and receive messages called tweets. The term twitter comes from the colloquial term which means "to utter successive chirping noises" or "to talk in a chattering fashion" (http://www.merriam-webster.com/dictionary/twitter). These are text-based posts of up to 140 characters displayed to an author's profile page and delivered to the author's subscribers (or followers). Users and followers can access Twitter by creating an account on the Twitter Web site: http://twitter.com/login. The U.S. Web traffic analysis site, Compete.com, estimated that as of February 2009 Twitter was ranked as the third most used social network with approximately 6 million unique monthly visits. Related capabilities include the ability to share images (e.g., cell phone pictures) and data via the Web site Twitpic (http://www.twitpic.com/), created in 2008 by Noah Everett.

Twitter users have effectively used this service to send eyewitness information for accidents, natural disasters, criminal activities, and terrorist events. Examples include:

- On May 12, 2008, the Chinese Sichuan earthquake was reported and tracked by Twitter reports.
- Twitter reports were used by local observers acting as "citizen journalists" [15, 16], as well as used by the terrorists in the Mumbai terrorist attack of November 2008.
- An eyewitness report of the U.S. Airways Flight 1549 emergency landing in the New York Hudson River in January 2009 was first reported and captured via a cell phone and shared via the Twitpic site.
- In January 2009, the progress and status of the Colorado Olde Stage Road fire were reported and tracked by local residents using Twitter.
- In June 2009, Twitter reports by Iranian presidential election protesters were used as a major information source by national networks when the Iranian government blocked most sources of external communication.
- Twitter reports of crime and information from first responders are available via http://blog.crimereports.com/tag/twitter/.

The Twitter infrastructure can be used for human reporting (the human acting as a soft sensor) or as an ad hoc network infrastructure for sending data from traditional sensors (either from a cell phone itself via a photo, video, or cell-based sensor). In this case the human can play the role of sensor platform, simply carrying the sensor(s) involved.

4.2 HUMANS AS SENSORS AND SENSOR PLATFORMS

Given these differences at the hardware levels, there are many limitations on the potential and optimal uses of cell phone–based sensors and their implied human hosts. First, status monitoring on a 24-hour basis is impractical, as in a motor vibration scenario. Second, hazardous conditions ranging from extreme ultraviolet radiation to noise cannot be readily tracked in this manner. Third, while cellular coverage is vast, it is not yet universal, which limits how remote locations, such as oil fields, might deploy the technology. Finally, the fact that people and their phones move is desirable in some instances, but a deal breaker in others.

At the same time, human-borne sensors present some unique capabilities:

- Metadata collection and disambiguation can sometimes be achieved at the moment of sensing rather than derived (or guessed at) later in processing.

- Humans can control many facets of sensor operation, whether through knobs or sliders, gestures [17], actuators like a camera shutter, aiming, or an on-off switch.
- Humans can improvise and adapt, standing on a picnic table to get a better angle for a photograph or other reading, for example.
- Humans can move to follow a moving target or to evade danger.
- Humans can take initiative, combining sensory inputs to make decisions about artifacts of potential importance: if a crashing noise (or a bird call) is heard nearby, appropriate sensing actions can be undertaken.
- Humans can ask questions.
- Humans can detect characteristics difficult to discern electromechanically: nervous habits like toe-tapping, oddities of gait (as with concealed weapons), or subtle facets of crowd behavior such as refusal to board an open bus or sudden crossing of a street with or against a traffic signal.

Experiences by U.S. soldiers in Iraq and Afghanistan, for example, point to the potential value of human observers [18]. Carey [18] related the example of a soldier who, despite a decoy, spotted the real IED encased in concrete 100 meters away from the decoy. The marine reported, "That block looks too symmetrical, too perfect." In a concept termed, "Every Soldier a Sensor," the U.S. Army has developed training methods to improve the situation awareness and reporting abilities of soldiers [19]. The aim is to "train soldiers to actively scan and observe their environment for details related to Commanders' Critical Information Requirements indicators and report or act in a concise and accurate manner" [19].

Such attention to context is also used by experienced police personnel. Research conducted at the Pennsylvania State University has involved interviews of patrol officers (by graduate student Alice Shapiro). Initial results indicate that officers are sensitive to anomalies such as nervous behavior or unusual emotional state (compared to surrounding individuals), movement by an individual within a crowd, or unusual location of objects (e.g., a backpack tethered to a light pole).

4.3 REPORTING MECHANISMS

Once a human equipped with a sensor has made an observation, he or she has many options through which to report it. Unlike traditional intelligence gathering, many of these mechanisms are overtly social in nature, raising the potential for both positive and negative effects. The open-source nature of these mechanisms raises further issues: the potential for spoofing, for example.

4.3.1 User-Generated Content

User-generated content is a blanket term that describes such tools as blogs, wikis, tagging, and uploaded photos and videos at sites such as Facebook, Flickr, and YouTube. The user-generated nature of this material means that it moves in a bottom-up fashion: people post things that strike their fancy, or things that they think might interest like-minded individuals. Most social media are intended for very small audiences of friends and family; production value (particularly in videos) can be extremely low. The equivalent holds true for print media; use of sources can be scrupulous or verge on the irresponsible.

Because social media are not generated with the rigorous standards of traditional information fusion in mind, accommodations must be made for their use in these venues. Even compared to newspapers, a traditional source of perhaps unintended insight into an area, social media can be extremely unreliable, and even when they are factually accurate, they may be packaged in unconventional ways.

All that said, several facets of social media commend it:

- Scale: The open collaboration model can generate extremely wide user participation. Flickr recently added its 4 billionth photograph in late 2009. Wikipedia includes over 3 million edited topics in English alone. YouTube is the fourth-busiest site on the Web; Facebook, MySpace, Blogger, and Wikipedia—each a social media site—rank in the top 10 [20].
- Coverage: Wikipedia posts material in over 100 languages. Flickr includes photographs from every continent, millions of them geotagged.
- Globality: Hi5, a top 20 site worldwide, supports 37 languages and is the #1 ranked social networking site in 30 different countries. QQ is a fast-growing social networking site in China. VKontakte has 20 million users in Russia, while MySpace includes 30 country/language options.

4.4 CHALLENGES AND BIASES IN GROUP OBSERVATIONS

Traditional data fusion can improve an extended sensor system's performance in four ways:

1. Representation: Parts can convey a sense of the whole.

2. Certainty: Good fusion practices increase the probability of a gathered piece of data.
3. Accuracy: Good fusion practices reduce noise and errors.
4. Completeness: Adding new knowledge to the current environment can fill blind or fuzzy spots in the view of that environment [21].

Given human beings carrying cell phone–based sensors, how do these criteria change? What will constitute fusion of data from sensors that can autonomously initiate coverage, move, improvise, and potentially get their human hosts killed?

4.5 TASKING THE COMMUNITY

In networks of sensors running on low power with limited processing capacity and finite bandwidth access to the immediate and greater environment, powerful algorithms have been necessary to reduce duplication, separate signal from noise, and coordinate multiple readings on a common event. Such algorithms, used on the back end of the process, will grow in importance as sensor types multiply, decision cycles shorten, and hostility toward sensors and their owners will likely intensify in information warfare and similar settings.

Humans in the loop complicate, augment, and sometimes obviate the use of these algorithms. A complex issue arises in tasking humans as sensors or humans with sensors. Robotic behavior may be in some ways desirable, but numerous issues preclude such strict tasking. Human initiative can be a powerful force, but channeling curiosity into defined information flows will be a challenge. Asking people only to carry a sensor in a random but geographically constrained pattern may have certain applications; asking people to report on activities of interest ("Follow whatever you think might be interesting"), or outside given parameters, has a different degree and kind of potential. In addition, data fusion has the potential to go from algorithmic to combining elements of both computational and social information processing.

Issues of bias in both tasking (elicitation) and reporting will be more complex than in electromechanical scenarios. The person carrying, aiming, or serving as the sensor may not understand or want to understand the individual delivering assignments. Even such words as "target" or "subject," when used in a surveillance context, convey powerful assumptions and possibly presumptions of guilt, innocence, or culpability. Both active bias and "passive" misunderstanding in either or both directions may introduce uncertainty.

A key development in wireless banking may be relevant to the incentive question. Cell phones are being used as currency storage and movement devices, particularly in Korea and parts of Africa [22]. The implications of monetary transfer to, or potentially from, the human sensor platform raise broad, complex questions.

The first of these questions are issues related to gaming the system: given sufficient incentives to do so, people will cheat or otherwise distort the original rules. Second, how will incentives be structured to protect privacy, safety, and other personal attributes? It is quite easy to envision an individual endangering himself or herself to get the more accurate data that would trigger a higher payment tier. Third, who pays? If the cost of running a human-powered (or human-augmented) sensor network can vary by orders of magnitude, what options can be put in place for the network owner to manage its expenses? New models will be required to trade off predicted (and realized) accuracy, cost, exposure, and robustness. Fourth, what sorts of competition issues may emerge between network members? The list goes on; data management of human sensor platforms breaks so much new ground that it will require textbooks devoted only to this one field.

4.5.1 Related Tasking Issues

Humans utilizing cell phone sensors can also be addressed before, during, or after sensing. Each of these time frames raises issues.

Before

In the interval between the human's being equipped with a sensor and a particular "mission," many things can change: the number, type, and availability of deployed sensors, the overall objective of which the sensors are one supporting element, and the context to be sensed. Having people who can change directions, understand updated instructions, and use human judgment presents some opportunities for near-real-time redeployment. At the same time, managing the entrepreneurialism of people is a departure from managing chipsets, batteries, and adhesion or other fastening technologies.

During

Given the high-resolution screens characteristic of increasingly more mobile phones, mobile devices are used to display bar codes that serve as discounts, authentication, and tracking devices. The sensor itself can be used as authentication for its bearer to enter a secure area, for example, and then as a beacon to ensure that the individual stays out of danger, classified activities, or another sensor's territory. Finally, humans can integrate information from multiple sources in the moment: if seismic sensors indicate activity, humans might activate either a camera to capture anticipated structural damage or a toxin sensor to capture evidence of gas leaks.

After

Simple questions such as "What was happening outside the camera's view?" or "What did you think was going on?" cannot be posed to traditional electromechanical or optical devices. People can also be debriefed to provide feedback on the sensing process at different layers of granularity. Such questions as "What tools do you wish you had? How hard was the unit to carry? Did the sensor slip into and out of its carrying case smoothly? What questions should we have been asking going in?" present a radically new scenario for process and product improvement.

4.6 INFORMATION MANAGEMENT

The collection, storage, and processing of sensor data depend on many factors. When each of these is considered in a cell phone model, new constraints and possibilities emerge.

4.6.1 Centralized Versus Localized Sensor Fusion

This long-running tension plays out in new ways in cell phone sensor networks. First of all, human involvement could potentially accelerate resolution of sensor input into data or information. Second, given the relatively high processing power of a contemporary smart phone (a reported 2.1 MIPS in the case of first generation iPhones), distributed signal processing can produce new kinds of results. Finally, the essential sensor-sink relationship will most likely need to be reinvented.

The localized/centralized issue relates directly to the issue of queries. To what extent can a human sensor network be queried, and to what extent will the sensing process be altered by explicit or implicit questions in the minds of the sensor bearers? How much data will human networks produce that fits poorly, if at all, in traditional sensor-friendly data stores? Perhaps most importantly, how can downstream analysts be equipped to investigate cell phone sensor data repositories when many of the captured attributes may not be captured (or capturable) in metadata? Many categories of human knowledge are difficult to access by keyword search, much less by SQL-like statements. When the sensor platform is (or is guided by) a human, the possibility for data of interest without adequate naming conventions increases.

4.6.2 Routing

While big batteries and high available bandwidth release certain constraints of classical sensor network routing, new elements factor into the equation. How does sensor traffic get treated—on the device, in the network, and at its destination—relative to human voice traffic, YouTube uploads or downloads, and other competing data streams? How will cellular, non-IP data traffic merge with traditional sensor feeds, WWW sources (e.g., weather), and human-generated sideband data that may be related to, but not included in, the sensor data? At what junctures do various facets become fused, and how irrevocably? How much access do data consumers have to raw feeds, or even first- and second-order summaries? A special issue in routing of cell phone information via a service such as Twitter involves the evolving pedigree of information: who observed what and when, what information (or misinformation) was added along the Twitter routing, what is the pedigree of the data and information, and related topics.

4.6.3 Representation of Uncertainty

As ad hoc reporting spreads to new populations, demographics, and geographies, multiple cultural norms, patterns of behavior, and incentive arrangements will come into play. Wherever the locus of analysis and decision-making, data from the community of nonprofessional observers will need to be characterized in its uncertainty, and the uncertainty will need to factor into subsequent uses of the data. Chapter 3 described some of the many issues involved in characterizing the observer (e.g., age, training, gender, focus of attention, level of training, mood, and interaction with other observers).

In the reporting process, uncertainty and errors are introduced by language. Forcing people to use standardized terms in support of a formal ontology will likely fail. In contrast, crowd-generated metadata, commonly known as tag clouds (see Figure 4.4), allow more commonly used labels to emerge while still preserving the breadth of original inputs. Outside of law enforcement, the military, or a professional community such as law or medicine, training practitioners in the use of a domain-specific vocabulary is impossible at scale. Use of computer-assisted knowledge elicitation (e.g., showing icons of potential targets or threats and the use of slider interface devices to quantify characteristics to a common scale) fail when new kinds of situations or activities are observed. Indeed, the very kinds of information that only humans can make (e.g., "that block looked too symmetrical") cannot generally be elicited by a specific form, set of a priori questions, or selectable icons. Instead we need to address the challenges of natural language to obtain the value of human observers.

An example of the complexity of related terms is shown in Figure 4.4, which is a "word cloud" generated by information annotated on images on Flickr. The

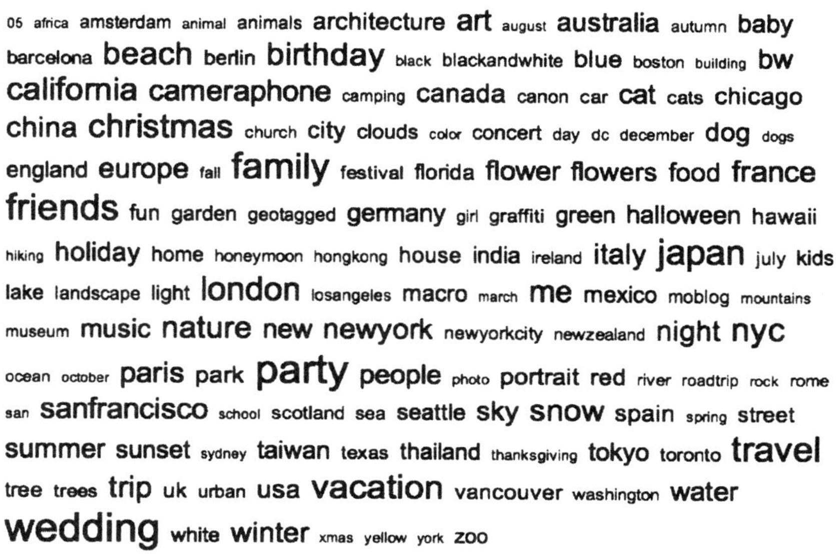

Figure 4.4 Tag cloud of photo-related labels from Flickr.

concept of creating an a priori ontology that describes or characterizes situations, activities, or events of interest may not be feasible. Instead, one may need to generate a "folksonomy," an ontology created by users via an evolutionary process, rather than a top-down a priori process.

As the Harvard Dialect survey discovered, even U.S. citizens can vary widely in their usage of language [23]:

Question 60: What do you call the area of grass between the sidewalk and the road?

a. berm (4.01%)
b. parking (1.75%)
c. tree lawn (1.92%)
d. terrace (0.73%)
e. curb strip (8.65%)
f. beltway (0.17%)
g. verge (2.56%)
h. I have no word for this (67.92%)
i. other (12.30%)

Pedigree is another potential axis for representing uncertainty. Much as Amazon.com lets customers comment on the helpfulness of product reviews, analysts may well tag the reports from a given source or group of sources as particularly timely, accurate, or insightful. The crowd itself already performs a similar task in many online media, highlighting contributors with strong track records. Similarly, members of the crowd may assist in rating sources, types of information and reports, and issues or problems—much like *Consumer Reports* obtains feedback from consumers to assist in rating products.

Calibration represents another possible method for representing uncertainty. Whether various observations are averaged, correlated, or otherwise used to assess performance, the sum of results in a given domain can generate quality metrics. An alternative calibration scheme might draw on known and potentially controlled benchmarks as a check on the performance of ad hoc observers in the detection and accurate recording of a given phenomenon.

4.7 CROWDS AS FUSION

We will consider such topics as prediction markets, wikis, and other forms of crowdsourcing in subsequent chapters. In the context of a "global neighborhood watch," however, it is becoming clear that groups of people can serve as powerful mechanisms for characterization, distillation, and other forms of information processing. A single example should suffice:

> On September 3, 2007, the record-holding aviator Steve Fossett disappeared during a flight over Nevada. In addition to Civil Air Patrol and other search efforts, the tech community utilized a technique that had been tried, unsuccessfully, to find a Microsoft researcher whose sailboat disappeared off California. Amazon has a service called Mechanical Turk (named for a chess-playing human inside an eighteenth-century automaton) that delivers Human Intelligence Tasks (HITs) from computers to humans. Image recognition is a common HIT, so satellite images of the area were captured and donated to the effort. Ten thousand volunteers were presented with satellite photos and asked to spot signs of life (trails in dirt or sand), pieces of wreckage, and the like. The technology was unsuccessful as Fossett's bones were found roughly a year later in mountainous terrain.

The principle of computers coordinating human image- and speech-recognition efforts holds great potential for traditional information fusion. As opposed to humans as sensors, humans as information processors can be linked more tightly to conventional mechanisms: Mechanical Turk was essentially a big human lightboard powered by volunteers. The advantage of asking multiple sets of eyes to score an image (think of a team of radiologists evaluating a mammogram), potentially scrambled to remove identifying characteristics or divorced from any context, should increase the quality of any assessments.

4.8 CONCLUSION

The projected growth of cell phone–based sensors will have many implications. It has the possibility to add new complexity to such already-difficult activities as sensor tasking, database design, multisensor (multiplatform?) data fusion, and security. This new model also has the potential to change the relationship of that which is sensed to the sensor if said sensor is a human being—perhaps in the service of an enemy of the state. Most positively, empowering people with sensors on a common, low-cost platform has implications for the practice of science and the conduct of society. If people can validate concerns about environmental toxins, reduce traffic for all commuters, or increase their awareness of their natural and man-made surroundings, broad sensor deployment has the potential to change how people relate to both their surroundings and their fellow humans [13]. However the story unfolds, the sensor community will find itself enmeshed in far broader and more public debates than ever before as the stakes get higher and the implications hit closer to home for those outside the community as traditionally constituted.

References

[1] Boisot, M., and B. McKelvey, "A Socio/Computational Method for Staying Ahead of Terrorist and Other Adversities," G. G. S. Suder, (ed.), *Corporate Strategies Under International Terrorism and Adversity,* Cheltenham, UK: Elgar, 2006, pp. 38–55.

[2] "Garbage Workers Help Keep City Safe," *Jacksonville News*, September 23, 2009, http://www.news4jax.com/news/21093406/detail.html.

[3] Phillips, A. L., "Of Sunflowers and Citizens," *American Scientist Magazine*, September-October, 2008, pp. 375–376.

[4] http://www.didyoufeelit.com/.

[5] Pitner, S., "Reporting News with a Cell Phone," Suite101.com blog, December 14, 2008, http://medialiteracy.suite101.com/article.cfm/reporting_news_with_a_cell_phone.

[6] http://www.ushahidi.com/work.

[7] "Halfway There," *The Economist,* May 29, 2008.

[8] Cusumano, M., and A. Gawer, *Platform Leadership: How Intel, Microsoft, and Cisco Drive Industry Innovation,* New York: Free Press, 2002, pp. 2–3.

[9] "Cell Phone Sensors Detect Radiation to Thwart Nuclear Terrorism," *ScienceDaily*, January 24, 2008, http://www.sciencedaily.com/releases/2008/01/080122154415.htm.

[10] California Center for Innovative Transportation, "Using GPS Mobile Phones as Traffic Sensors: A Field Experiment," http://www.calccit.org/projects/GPS-Mobile-Phones-as-Traffic-Sensors.html.

[11] Nokia Eco Sensor Concept, http://www.nokia.com/A4707477.

[12] Blum Center for Developing Economies, Telemicroscopy for Disease Diagnosis, http://blumcenter.berkeley.edu/.

[13] Noguchi, Y., "Camera Phones Lend Immediacy to Images of Disaster," *Washington Post*, July 8, 2005.

[14] Hansell, S., "Have Camera Phone? Yahoo and Reuters Want You to Work for Their News Service," *New York Times,* December 4, 2006.

[15] Hughes, A. L,. and L. Palen, "Twitter Adoption and Use in Mass Convergence and Emergency Events," *Proc. of the 6th International ISCRAM Conference*, Gothenburg, Sweden, 2009.

[16] http://www.forbes.com/2008/11/28/mumbai-twitter-sms-tech-internet-cx_bc_kn_1128mumbai. html?feed=rss_news.

[17] Rohs, M., and P. Zweifel, "A Conceptual Framework for Camera Phone-Based Interaction Techniques," in Gellerson, Want, and Schmidt, (eds.), *Pervasive Computing: Third International Conference, PERVASIVE 2005*, Berlin: Springer, 2005, pp. 171–189.

[18] Carey, B., "In Battle, Hunches Prove to Be Valuable Assets," *New York Times*, July 28, 2009, p. 1.

[19] http://www.peostri.army.mil/PAO/pressrelease/SLIM_ES3.jsp.

[20] Alexa.com.

[21] Bellot, D., A Boyer, and F. Charpillet, "A New Definition of Qualified Gain in a Data Fusion Process: Application to Telemedicine," *Proceedings of the Fifth International Conference on Information Fusion,* Annapolis, MD, 2002.

[22] "The Future of Money," *The Economist,* February 15, 2007.

[23] Harvard Dialect Survey, now hosted at the University of Wisconsin-Milwaukee, http://www4.uwm.edu/FLL/linguistics/dialect/staticmaps/q_60.html.

Chapter 5

What Does It Mean to Live in a Searchable World?

"The web is where society keeps the sum total of human knowledge. It's where we learn and play, shop and do business, keep up with old friends and meet new ones."

—I. H. Witten, M. Gori, and T. Numerico, *Web Dragons: Inside the Myths of Search Engine Technology*, Morgan Kaufmann Publishers, Elsevier, 2007, p. xv

5.1 INTRODUCTION

While widespread availability of search technology is only about 15 years old, the implications continue to ripple outward. Work, commerce, medical care, mate-finding, crime, and education all are being reshaped by an effectively infinite base of information made usable by various types of search, indexing, and related technologies. Human-centered fusion would be a much more constrained exercise if search, along with mobile telephony and data platforms, had not reshaped the human landscape in such a brief window of time.

The creation and evolution of the Web and associated search capabilities provide major opportunities and changes to traditional information fusion. Historically, data fusion systems referred to linking one or more sensors together via a local area network and combining the sensor data via a wide variety of estimation, pattern recognition, and automated reasoning algorithms. However,

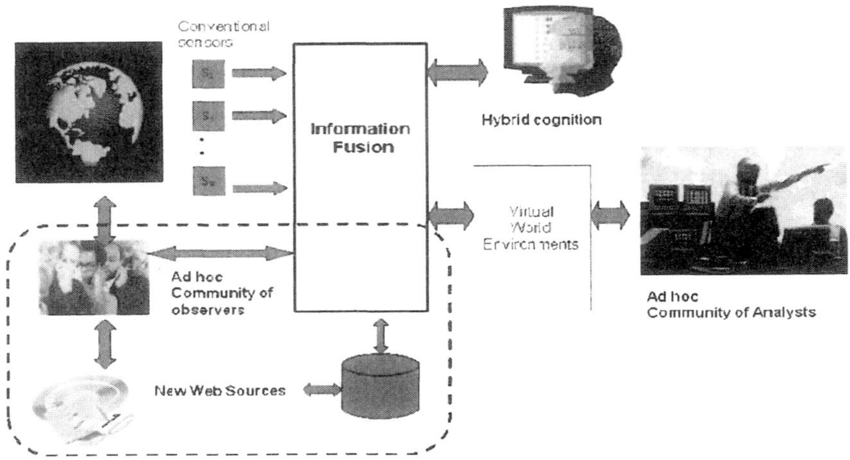

Figure 5.1 New sources of information for human-centered fusion.

these traditional sources of information are greatly augmented by information available on the Web. Figure 5.1 shows that new fusion systems may access traditional sensor data (shown on the top left side of the figure) with data collected by ad hoc human observers and information on the Web (shown in the dashed line on the bottom left side of the figure). Data available on the Web may include human observations and comments (e.g., via blogs, micro-blogs, and news reports), as well as online access to sensor data (e.g., via webcams and Internet connected sensor systems) and other data such as Flickr images. Extending the concept of fusion to include these new sources of information provides new challenges and opportunities for information fusion, ranging from how to determine what data or sources exist to tasking (e.g., via search engines or information requests to ad hoc observers) to source characterization, determination of source pedigree, and other issues. Comments on the impact of search capabilities on human-centered fusion are provided in Section 5.5.

5.1.1 A Brief Historical Perspective

For the digital natives [1] who have always known the Internet and surfing the Web, it is difficult to understand that the Internet is a relatively new phenomenon. We note that while the terms *Internet* and *World Wide Web* (or simply *Web*) are often used interchangeably, technically the Internet is a networking protocol that enables physical communications among computers connected to a network, while the World Wide Web is a software protocol that allows users to access files stored on Internet computers [2]. The development of the initial physical infrastructure for the Internet (called ARPANET) was funded by the U.S. Department of Defense Advanced Research Projects Agency (ARPA) and performed by BBN Corporation, which, in 1969, established four nodes at:

1. UCLA;
2. The Stanford Research Institute;
3. The University of California at Santa Barbara (UCSB);
4. The University of Utah [3].

This infrastructure was originally developed to support scientific collaboration. In 1971, Ray Tomlinson of BBN invented an e-mail program to send messages across the distributed network. Surprisingly, it required another 7 years for the first spam e-mail to be sent by a DEC computer company marketing representative as a new product announcement (starting a trend that has resulted in the current situation in which over 95% of all e-mail is spam).

During the 1980s, international connections were developed, and by 1989 the number of Internet host computers exceeded 100,000 machines. As of mid-2006, the number of computers acting as Internet hosts was estimated to have exceeded 439 million [3]. Despite the rapid increases in physical connectivity, the general public use of information on the Internet awaited development of tools for ready search and access. Tim Berners-Lee, a consultant at CERN (the European particle physics laboratory in Geneva), developed a prototype hypertext editor, and in 1991 sent a message to an Internet newsgroup announcing the availability of the tool and provided a link for its free download. A year later, in 1992, the World Wide Web was demonstrated and distributed along with the browser software [4].

A history of search engines is provided by [5]. The first tool for actual searching on the Internet was called Archie (developed by a student, Alan Emtage at McGill University in Montreal), and a search engine for plaintext documents called Gopher (named after the University of Minnesota mascot) was created in 1991 by Mark McCahil. The mid- to late 1990s witnessed the creation of numerous commercial search engines such as Yahoo!, Lycos, and AltaVista (all introduced in 1994). Google was introduced in 1997, and more recently Microsoft launched Bing in 2009 as an intended competitor to Google. Various attempts have been made to create more friendly interfaces to search engines, including the use of natural language questions introduced by AskJeeves (now www.Ask.com), with mixed success.

5.1.2 Intellectual Earthquakes

It remains to be seen how the search era will be positioned in the grand sweep of human intellectual history. In the short term, five interrelated developments are noted.

1. *Ubiquitous access:* Knowledge has become virtualized in that vast quantities of, but far from all, information have been separated from location. Not long ago, national and university research libraries required decades or centuries to assemble, vast sums of money to acquire and maintain, and large staffs of specialists in often arcane disciplines such as cataloging, book-binding, acquisitions, and archival management. Beginning with text-string matching and extending through link analysis, metadata matching, and semantic analysis, technology has made available billions of pages of open information. Searchers can be anywhere and in any number: the virtual resource can be used by as many people as need it at a given instant. Moreover, data from university and government laboratories can be accessed remotely without traveling to, or directly interacting with, the physical laboratory or sensing instruments. Examples include interactive weather satellite images from the NASA Earth Science Office (http://www.ghcc.msfc.nasa.gov/GOES), multiple Web-based laboratories for Internet remote experimentation [6], virtual telescopes (http://skyview.gsfc.nasa.gov), and worldwide access to webcams (see http://www.earthcam.com/).

2. *Virtualized knowledge:* Knowledge has also become virtualized in that books and paper journals no longer retain unquestioned primacy as the mechanisms for knowledge dissemination. Bloggers, e-mailers, or video-recorded statements can find a worldwide audience in a matter of hours as opposed to months or years. In addition, the cataloging systems for books and journals no longer can stand as authoritative taxonomies of the state of a field.

3. *Democratization:* Knowledge has been democratized, as many of the resources formerly available in those rare physical repositories noted above, including search engines themselves, are now ubiquitous. LexisNexis licenses, while unquestionably valuable, cost in the tens of thousands of dollars (one published figure quotes $300 per hour): Google delivers search results for a few tenths of a cent that cost users nothing and advertisers only cents.

4. *New search skills:* While information is liberated from the limits of print, search results frequently lack context. Formulating searches has become a new skill that can shape the quality and relevance of one's results.

Knowing when to dig deeper, when to jump sideways, when to retreat and reformulate, and where to look still requires sophistication beyond the simplicity implied by the clean input screen. In addition, results are not objective, yet the ranked pages of results imply an ordering of weight or fit that is difficult to consciously override. In prior times, a researcher might consult with a professional librarian to understand the intricacies of the Dewey Decimal System. Such interactions are replaced by utilization of specialized sites such as www.ask.com.

5. *The primacy of search tools:* Search is perhaps the tool for the times, connected to related developments.

    a. While it has become a truism to talk about an "information economy," Stanford's Paul Romer introduced an academic formulation in the late 1980s and early 1990s. Commonly called new growth theory, Romer's schema asserts that land, labor, and capital no longer constitute the basis of a modern economy. Instead, ideas create a significant proportion of economic value. For example, conventional economics would see the end of the oil era as portending the end of transportation. In Romer's view, however, human ingenuity will find more fuel sources and economically viable technologies for utilizing them given the market rewards for doing so. The point for our purposes is that search becomes an essential utility for making the "information economy" run [7].

    b. In an information economy, certain kinds of work and value creation become dissociated from their consumption. A nurse must be in the same room as a patient to give an injection, but a radiologist or a teacher can be half a world away. This is the main thesis of Thomas Friedman's familiar characterization that "the world is flat" [8]. Given the democratizing forces noted earlier, search is part and parcel of the flattening process, making formal and informal knowledge accessible to anyone (who has a connection) at any time.

    c. Not only has information been dissociated from paper, but information production has also been even further removed from consumption. What Chris Anderson referred to as "the long tail" [9] and Nassim Nicholas Taleb explored in *The Black Swan* [10] both relate to a world in which power-law distributions explain increasing portions of phenomena. In a physical world, distributions of height, for example, fall on a Gaussian distribution and all eventualities are within one order of magnitude: no adult human is less than 1 foot tall and no one is more than 10 feet tall. In information landscapes, nowhere more obviously than the World Wide Web, billions of page views are allocated on a "fat-tail" distribution in which perhaps 1% of Web pages receive perhaps half of all traffic [11]. The remaining half, however, is spread over millions of sites, some with tiny audiences, the so-called long tail. In the process of navigating online bookstores or music services, search becomes essential: there would be no long tail without search and no need for search if not for long tails.

5.2 CONTEXT

Any piece of information has at least one, and probably several, relevant contexts. The rise of search has prompted the rise of multiple organizational concepts, some of which vary from traditional ontological systems in their bottom-up, "good-enough" character.

5.2.1 Perspectives on Information Collection and Access

From the time of writing, assembling information has conferred benefits, whether of strategic or tactical advantage, prestige, or erudition. Given the high value of books, many were chained in early Greek libraries and later libraries such as the Sorbonne [12]. Later, the advent of the public circulating library was intimately connected to considerations of political organization: Benjamin Franklin's role in both library development and the American Revolution separates him from many political figures both before and after. Over time, the emphasis gradually shifted from assembly (as at Alexandria) to classification, most notably among the French encyclopedists but also in the Library of Congress system. In the digital age, assembly (in the form of the Internet Archive) has fallen behind the success of search, which renders both classification and assembly less relevant.

In the early 2000s, search has facilitated a widely decentralized production of information, both printed and visual. Finding one's way among that growing volume of data and information will hinge on how three current forces play out going forward:

1. *The growth of hidden data:* The growth of online but hidden databases (behind a search screen and firewall, for example) known as the "deep Web" means that search engine crawls reveal only a tiny percentage of online information [2]. If information elements are custom-built in response to a query, for example, the resulting page will not be indexed [13]. Federating searches and/or crawls remains controversial, in part because of allocation of advertising revenues. In addition, deep Web information is often proprietary and is held close for competitive reasons; airline flight schedules and product catalogs or price lists are some examples. Other examples include

performance data and failure rates of machinery, industrial processes, and even medical procedures.

2. *Generation of metadata:* Finding information typically relies on metadata, which is being handled in two broadly defined ways. From the top-down perspective, semantics (systems of meaning) are being built to coordinate data, particularly for machine-to-machine transactions. Formal semantic maps, known as ontologies, are extensive, labor-intensive, and rigid. In some circumstances they are essential; in others, they are little more than a nuisance, particularly if they are implemented but not maintained. At the scale of the World Wide Web, several efforts are under way to utilize semantics to enable database-like queries as opposed to text-string–based searches. Barack Obama attended Columbia University and later was elected president. To ask "How many U.S. presidents attended Columbia University?" of a search engine would fail, but to query a database of U.S. presidents would be trivial. Such services as Metaweb's Freebase and Radar Networks' Twine are attempting to organize information with sufficient disambiguation and classification to make such queries possible. Other efforts focus on the process of scientific publishing, where a working vocabulary is potentially easier to define and organize.
3. *Human assisted tagging:* From the bottom up, tagging is the practice of site visitors attaching metadata based on a personal view. No effort is made to reconcile conflicting terminology; instead, simple visualizations show popularity of various tags. Particularly for image data but also text in the absence of an abstract, tags can provide cost-effective first approximations of meaning: they answer the simple question, "What is this picture, video, or blog post about?" [14]

Each of these three developments contributes to a significant area of impact for human-centered fusion. For one, fusion can happen at a high level of technical capability and on an extremely broad base of raw data outside traditional fusion channels. For another, good-enough open-source efforts can improve decision-making for anyone, including one's adversaries. Finally, commercial search and other technologies can become sufficiently powerful that they merit inclusion in formal fusion processes, whether human centered or otherwise, in part because of the proven scalability of such tools as MapReduce and other techniques made common by search [15].

5.2.2 From Classification (Ontologies) to Networks (Indexing)

For the moment, search technologies and business models favor the vast scalability of text-string matching with no attempt to impose order. Recall that Yahoo! originally began as an index, a sort of Internet Yellow Pages, before the human-powered model failed to keep pace with the vast size of the Web. Currently, search is the lowest-cost, most scalable approach to finding (rather than organizing) things that users think they want to find. While efforts to build ontologies will by necessity continue in relevant domains, the economies of scale that accompany the crawl-index-retrieve model make it the dominant approach.

Significant exceptions to this rule exist in the realm of social networks, where asking people something (that often has particular meaning, such as "best") is often a faster way to a satisfactory answer. Given the scale of such networks as LinkedIn or Facebook (with over 250 million users by mid-2009), the odds are that somebody knows somebody that knows the answer to a vast number of potential queries that would fare poorly in a search engine. More generally, a key tenet of the so-called Web 2.0 is that groups can achieve substantial tasks given what technical publisher Tim O'Reilly calls an appropriate "architecture of participation" [16]. This might include tagging, recommendations, search queries, or other mechanisms for community input, many of which can improve search processes and results.

5.3 DOMAINS OF CHANGE

Search and the World Wide Web are in many ways joined at the hip: growth in one requires, and facilitates, growth in the other. Four domains of change—chosen from a much longer list—begin to illustrate the sweep of the ways that search is altering personal life, commerce, and other aspects of the world.

5.3.1 Intellectual Life

As we have discussed, the role of books, librarians, and professional journals is changing with the greater availability of online resources. Given that the marginal cost of putting an idea into circulation approaches zero for access to a potentially worldwide audience, the rise of a new kind of public intellectual coincides with the introduction of easy-to-use blogging and later microblogging technologies in the past decade. Any intellectual niche, and any perspective, whether defensible, vetted, or not, can find an audience. Bypassing the gatekeeper has many implications for public discussion, not to mention public safety and political stability if the information is inflammatory, undocumented, or just plain wrong.

Truly rigorous inquiry, in sciences both hard and social, arts, engineering, or the humanities, can be difficult in its long timetables, expensive infrastructure, or contention with dominant theories. Democratized publishing can be more nimble, but particularly in blog postings, the error rate can be worryingly high. Search may be improving, but the percentage of bad information that can be retrieved cannot be dismissed. Mechanisms for validation can be difficult within search rankings that factor in link structure and traffic analysis. A whole cluster of mutually reinforcing pages may all score well and reinforce each other's point of view, but in the "blogosphere" particularly, echo chambers are a very real phenomenon. Information, in short, can be both wrong and reinforced by a network of other sites with bad information.

The opposite side of this opening up of intellectual life is more encouraging. As the marginal cost of "looking it up" approaches zero, new habits are taking shape. Whereas a personal reference library costs thousands of dollars to acquire and update, anyone now has access to a vast fund of information, much of it credible. The need to guesstimate or factor in error margins has been reduced for many categories of discussion. With a little cleverness, amateurs are assembling truly enlightening information exercises, whether mashups of multiple, possibly unrelated data sources or creating ad hoc sites to support the public services of neighborhood watchdog activities, or resource assembly. With the end of large capital requirements for information dispersal as the broadcast model is joined by "pull" modes of access, amateurs can do and are often doing valuable things. Examples range from pothole mapping to house price maps with mass transit commute times overlaid to an application for the iPhone that tells moviegoers what scenes to skip for bathroom or popcorn runs.

Making the vast resources on the Web available to a distributed audience, largely via search, facilitates new kinds of peer review. The Linux operating system [17] and Wikipedia are two widely cited examples, but the journal *Nature* discontinued an experiment in open-source peer review in 2006 after minimal participation [18]. Academic publishing, meanwhile, continues to be challenged by alternatives to the current model with its particular mix of strengths (prevention of many kinds of errors, preservation of prevailing constructs) and weaknesses (long duration between finding and dissemination, closed communities, anonymity that can cut both ways, high barriers to innovation).

As newspaper and magazine publishers are seeing in their readers' behavior, reading and writing are both different in the aftermath of broadly available search. Such tricks as slideshow features are designed only to drive page views for ad rankings, while readers and bloggers often "deep-link" to pages that follow the beginning of an article, potentially bypassing important context while also weeding through to one relevant assertion within a longer body of material.

The immediate gratification of a decontextualized info-nugget, however, must be counterbalanced against the need for sophisticated, nuanced points of view to build over time, acknowledging contending perspectives and often facts that do not fit the argument. Particularly because it does not pretend to understand meaning, current search technology may not be doing as much as it might to advance the state of knowledge. Finding the nearest pizzeria is valuable and relatively uncontroversial, but many other questions fail to map to a text-string-based paradigm. Some examples might be "Where should I book a vacation house?" or "What are the benefits and liabilities of a protectionist trade policy?" The state of North American public discourse is exhibiting some signs of polarization that might in part be rooted in the decontextualization and "nuggetization" of complex ideas and concepts both online and in other media.

Finally, search's change in intellectual life pertains to education, learning, and training. Just as the invention of the pocket calculator spurred a debate that is still not entirely settled over the place of mastering methods versus memorizing content, search opens vast information resources—and makes them amenable to cutting and pasting. One example involves the creation of "instant bibliographies" on research topics by using tools such as EasyBib (http://www.easybib.com/) which allows multiple users to access bibliographies created by other researchers (and archives evolving bibliographies from collective users). Similarly, Web sites such as Delicious (http://delicious.com/) provide a means for social network sharing of Web bookmarks; as a user searches for reference Web sites, the Delicious site recommends similar sites based on the identified preferences of the collective users.

Searchable information means that institutional and public memory need not be locked in city halls and library basements, yet something is lost when states and capitals are no longer memorized, for example. More broadly, both education and training practitioners face the question of what skills need to be taught both to capitalize on and to compensate for the shape search imposes on a citizen's world.

5.3.2 Commerce

The changes being brought to the world of commerce by search cannot be enumerated here. The following examples are meant to be suggestive rather than exhaustive.

News and Media

Google is reinventing the advertising model, from a statistical approximation of audience to a highly instrumented system in which advertisers pay not for abstract "reach" but concrete user behavior: click-throughs, return visits, and purchases. Yahoo!, meanwhile, hired a CEO from Hollywood, but it was the technologists at Google that won the day in media reinvention. The newspaper industry, in the form of a set of recommendations from the American Press Institute, referred repeatedly to Google as a central player in the decline of the conventional paper-based business model [19]. The *New York Times* parent company is selling off assets in the face of a steeply declining share price, while the *Chicago Tribune*'s parent company filed for bankruptcy the same day that the paper broke the story of the arrest on corruption charges of the state's governor.

Secondary Markets

Services such as eBay and Craigslist have turned secondary markets for used and even new but discounted goods into global arenas. Finding a valuable antique at a yard sale is much more difficult given market liquidity and price transparency. Used cars have moved from being primarily a local transaction to gain the scale and reach of national and even international markets, again through eBay Motors and similar services. The primary interface to these long-tail markets is search, whether delimited by price, model year, proximity, or the vehicle's features. Even industrial surplus and capital equipment auctions are being augmented by online presence: an aircraft carrier appeared on eBay in 2004, to take one extreme example.

Employment and Work

The process of job-hunting has been thoroughly changed by search, in both directions. Applicants' dossiers include search results as well as social network artifacts, while applicants often try to navigate around the human resource (HR) screening process by using search tools to find the hiring executive directly. Intelligence-gathering includes extensive collections of insider and third-party opinions that rarely reinforce the imagery conveyed by the prospective employer's public presence. Once again, geography is eliminated as a barrier: finding all current openings for a specific skill in any or all states becomes a matter of keystrokes rather than of obtaining as many (physical) Sunday newspapers as possible, often the only alternative as recently as the 1990s.

Automobiles

The widely lamented asymmetry of information between buyer and seller, made worse with used vehicles, is in flux given the expansion of search capabilities. Accident and recall reports for a specific Vehicle Identification Number (VIN) are readily available, as are some service records. User experiences with a particular dealer, vehicle, or feature are shared widely. Multiple dealers can be addressed by a single inquiry; inventory lists are often visible, as are invoice prices, rebates, dealer programs, and other formerly proprietary information. As the entire industry continues its process of reinvention, dealerships, pricing, marketing campaigns, and perhaps even vehicles themselves (given the evolution of navigation systems, which are a near neighbor to search) will look less and less like their 1995 forerunners.

Real Estate

Like automobiles, real estate is an expensive, considered purchase, making it ripe for heavy utilization of search capabilities. Indeed, the National Association of Realtors has mandated industry data standards such that essentially every house offered by a Realtor nationwide can be searched from anywhere. Such information as square footage, bedroom and bathroom counts, and exterior material are easily described. At the same time, however, real estate illustrates the limitations of search. The many intangibles involved in a residence and in a real estate transaction still almost always require nonstandardized processes and human intermediation. Such questions as "are the neighbors nice," "does the house have good light," and "is the street noisy" are typically impossible to search with any degree of confidence.

Retail

As Amazon illustrates as well as any company, search in retail is but one element in a rich information environment that also extends to tagging, user reviews (and reviews of reviews), collaborative filtering ("people who ordered this item also ordered these other items"), and institutional memory: "the last time Jane Doe released a book, you ordered three copies, and now she just published the sequel." Thus, search, while powerful in both attracting customers and in distracting them to a lower price, closer location, or more attractive offer, cannot be seen in isolation, most notably in relation to social media, which we will discuss later.

Travel

The demise of the old-fashioned travel agents, who were paid primarily on the basis of their ability to print authorized tickets, is essentially complete. Local agencies typically must find a specialty, such as cruises, ecotourism, or a specific region, while the national providers have consolidated. Carlson Wagonlit now provides travel management (vendor consolidation, internal policies and controls on travel spending, special event coordination) rather than pure ticketing. Rosenbluth was acquired by American Express in 2003. In a situation unlike real estate, in which a few airlines serve known destinations at known scheduled times, or hotel chains have properties at easily visualized locations, search rapidly reduced the value added by an intermediary.

Entertainment

Until the early twentieth century, audiences heard music only in live venues: an individual might well hear "Beethoven's Fifth" or "The Four Seasons" only once in a lifetime. With recording, music could be played whenever one wished, wherever there was a player, giving birth to an entirely new industry, based in part on the petrochemical innovations of the late nineteenth-century expansion of the chemical sector in Germany and then elsewhere. As recently as the 1970s, motion pictures would come through the local theater, giving audiences perhaps their only chance to see it. Secondary screenings sometimes happened via television, at college film festivals, and in other venues, but they were not guaranteed. Then the videocassette recorder, and more recently the DVD, made the world's cinematic archive accessible in the same way music was. In our time, television was primarily ephemeral: barring a few series with enough of a fan base to support a VHS collection, or individuals who taped broadcasts, seeing a rebroadcast of a specific event was often impossible. Now, search-powered video distribution via the Internet is turning the ephemeral into the convenient. If Thursdays at 9:00 p.m. is somehow inconvenient, both *The Office* and *Grey's Anatomy* are available on demand.

Furthermore, the rise of public information sources such as the Internet Movie Database (IMDB, owned by Amazon) facilitates domain-specific search, allowing fans to track a favorite actor's or director's career, identify incongruities in the final cut, compare industry award lists, and otherwise master a large body of material relevant to their interests. On Metaweb's Freebase service, for example, a demonstration query goes as follows: "How many movies has Jennifer Connelly appeared in opposite actors who had previously been directed by Steven Spielberg?" The answer can be found at: (http://www.freebase.com/view/guid/9202a8c04000641f8000000005b85930). Clearly, search changes the experience of being a fan.

5.3.3 Personal Empowerment

Even before AltaVista, Excite, or Lycos, people searched for information: directory assistance telephone calls were the most common form, but reference librarians were (and remain) an essential resource. Many searches now can be answered either directly, by the manufacturer, for example, but quite often by a member of a diffuse community who had a similar issue, opportunity, or experience. Five broad categories will be sketched out, once again, only suggestively rather than in any depth.

1. *Everyday needs:* While it lacks an elegant title, one category for which search is essential concerns one's everyday needs. How do I get the printer to work with the new computer? How can I get chocolate stains out of linen pants? What does that grinding noise in the car mean? Answers to such queries may be short or involved, be generated from databases or human (online) conversation, and save a little time or be relatively important, as in a safety recall for a transmission defect, for example. In this same vein,

social networks, while not a direct component of search per se (though Google owns Orkut, a highly visible social network site in Brazil and India especially), are one step removed in that similar queries can readily be directed to one's network, particularly via a microblogging service such as Twitter: "At liquor store. Serving chicken parm tonight and need a good $10 red. Any suggestions?" should be understood in the same domain as search.

2. *Navigation:* A major component of everyday existence is navigation, both in the literal wayfinding sense and more generally in getting oriented to an unfamiliar domain. Search clearly has changed the mapping industry, as the near-demise of CD-ROM road maps makes clear. Other processes, such as traveling abroad for the first time, buying a puppy, trying not to get ripped off by a mechanic or contractor, or choosing a party venue, lend themselves extremely well to search-driven results in that other people have the same question, there are often a finite number of right (or plausible) answers, and the keywords can be easily disambiguated.

3. *Dating and friendship:* Dating and friendship have obviously been revolutionized by search. Whether it is merely Googling one's blind date to check for red flags or filling in extensive (and expensive) survey instruments for matching engines at eHarmony or Match.com, search queries related to people's social and romantic pursuits have become a billion-dollar industry as of this writing. Needless to say, social networks play a crucial role here as well in both search and discovery processes.

4. *Family matters:* In addition to dating and relationships, it is not surprising that family matters lend themselves to search. Whether it is childrearing, dealing with aging parents, marriage advice, or genealogy, rich resources (along with other kinds) are readily available. This wide availability can supplement or supplant the role of traditional authorities such as churches, village elders, or human services professionals. The relevant sources can be definitive (an immigration record for a grandparent, a statute, or an application form) or tend more toward folk wisdom: here's what worked for me, this was my experience with this firm/institution/individual, or simply opinion, informed or not.

5. *Health and well-being:* The changes to health and well-being that relate to search are rapidly unfolding. Alternative medicine, a $34 billion industry by itself, obviously benefits from the leverage of online information versus the big pharma advertising model, for example [20]. The interaction between doctor and patient is evolving as patients come to office visits armed with extensive statistics from deep research that a generalist physician cannot duplicate: the center of authority in the relationship shifts [21]. At the same time, patients are banding together into support networks,

yet again reaffirming the connection between search technology and social networks: patients look up their disease, find relevant information contributed by other patients, and then potentially join the network themselves, enriching the source of knowledge and experience for the next searcher. In addition, the dissociation of knowledge from location mentioned earlier facilitates virtual disease communities that would be impossible for many people who have limited mobility or simply live far from other people who might have the same condition.

Given the broad democratization of information access, we come to a situation in which "every man [is] a knowledge worker," with due apologies to Huey Long and Peter Drucker. Broad questions are still unfolding: What constitutes literacy in a world where millions of facts are available at a keystroke? What are the rights and responsibilities of membership in any number of communities, up to, and including, citizenship? What digital divides will emerge between well-informed (and suitably discerning) people, richly ill-informed people who believe things they found on the Internet, and those who are not part of the conversation whatsoever? When enlisting citizens in human-centered fusion, what assumptions can and cannot be made about the "man on the street"?

5.3.4 Crime and Terror

An obvious downside of broadly searchable information is, of course, the ability for motivations, techniques, and context to be shared across the world. While it would be problematic to ask a reference librarian, or the feed store clerk, how to make a fertilizer bomb, one can now find multiple recipes in an instant. Whether it involves smuggling drugs into prisons, sniper positioning and training, or spoofing air-defense radar, answers are readily available.

Apart from the many practical matters of proliferation of techniques (to high school students, for example), there is the matter of what might be called accessible deviance. What does it mean when *any* viewpoint or practice can find adherents in, initially, a virtual world? People who formerly were ostracized for their beliefs (the Unabomber comes to mind) now can recruit flocks of loners, deviants, or troublemakers, giving them power in numbers.

At the level of criminality, the life cycle from getting into trouble, getting caught, getting incarcerated, and getting released now has many more variations because of the amplifying power of search: obtaining the paper-based resources or word of mouth to master any of these phases was considerably more difficult just a decade or two ago. Upon release, registration as a sex offender was one kind of phenomenon in a paper-based world but something entirely different when a nationwide list of offenders can be summoned up in less than a second. Balancing the rights of citizens with the right of a freed criminal who has completed his or her sentence is never simple, but it has recently become much more difficult.

Warfighting at multiple levels is in an obvious stage of transition. Search can reveal anything from the proper arrangement of service ribbons to details on weapons and infrastructure. Doctrine, strategy, and tactics are readily available. Getting involved in "information warfare," in many of its different meanings, is readily facilitated by code libraries, root kits, and other tools that have no analog in hardware weaponry. If asymmetry is premised on finding points of vulnerability and exploiting them, having anonymous access to everything from train schedules to maps to first-person accounts to toxicity information to arcane product catalogs is bound to increase effectiveness.

5.3.5 Implications of Far-Reaching Change for Human-Centered Fusion

In the face of search, GPS and related navigation technologies, and other tools for knowledge generation and organization, the discipline of information fusion is being forced to evolve. The context for any given interaction, whether human or mechanical, adversarial or routine, or commercial or personal, can no longer be assumed. Banking occurs far from banks, research into virtually any topic is no longer correlated with physical libraries, and people with similar search interests can now find each other. Interest groups can form in ways that would have been impossible 15 years ago. Personal empowerment can be taken to extremes if one is searching for weapons, tactics, or inspiration. Human participants in a formal fusion process will likely be participating in parallel in informal fusion efforts, whether by voting on a product rating, reviewing a movie, or tagging a vacation photo. Understanding and managing these various strands of collective human effort to collect and understand experience will constitute one new challenge for fusion experts; others will be discussed next.

5.4 LOOKING AHEAD

The rapid pace of technology innovation is expanding the capabilities of search to include other domains, such as sound and images. At the same time, the technologies have far-reaching implications for human issues of attention, civil liberties, and other constraints.

5.4.1 Where Is the Technology Headed?

Six broad and sometimes overlapping technology directions bear watching in the rapid evolution of this suddenly indispensible technology.

Video Search

The rapid increase of video traffic on the Internet for entertainment, education, and surveillance purposes is rapidly generating substantial data volumes that are essentially worthless without search. When dialogue and narration are involved, promising tools utilize language-independent syllable matching (as at Truveo, now owned by AOL), but image recognition ("find me all the people in red ski jackets who entered the train station in the last 12 hours") is still in the early stages of development.

Image Search

Along similar lines for still images, the tagging approach utilized by Flickr and similar services has outperformed technological tools for pornography filters, image search, and automated characterization. To date, people are much more effective at identifying even simple visual characteristics than are computers, meaning that machine-generated metadata, a prerequisite for large-scale image search, stands as a barrier to wider adoption. Extensive ongoing research is focused on computer automation of tagging or semantic labeling, especially for image data [22, 23], with varying degrees of success.

Semantics, Tagging, and Queries

Semantic tools for labeling and categorizing text-based input (into a data warehouse, for example) have entered commercialization, allowing for the possibility of exploratory search: visual and other forms of navigating large data sets without knowing a text-based query to begin the process. Speaking generally, the better the processing and indexing on the information at the loading state, the less the person needs to know on the output state. The Jennifer Connelly query noted earlier is impossible given the simple processing done by a standard search engine, which does not know that Jennifer Connelly is an actress, that actresses appear in movies alongside other actors, and that movies are a form of entertainment that require producers and directors. Machine-based semantic tools such as Expressor's Repositor product may prove useful in the enterprise model, although the human-powered tagging approach is scaling well for selected topic areas in both social Web public sites and beta projects inside several organizations.

Domain-Specific (Vertical) Search

In both research projects and start-up companies, limiting the scope of a body of knowledge increases the quality of search results even without semantics, although some of these efforts are built on vertically delimited ontologies. Commercial examples include Kayak.com for travel and novoseek in biomedicine. In university research, verticalization is most common in medical and scientific fields where the search engine may be accompanied by a tool bench of other applications such as maps, calculators, glossaries, and so on.

Overlap of Social Networking and Search

In scholarly fields particularly, social networks and knowledge production are deeply connected: the very concept of Erdos numbers predated Kevin Bacon and his six degrees of separation, and the processes of graduate student apprenticeship and coauthoring continue to reflect the ways that knowledge often originates in networks. Among knowledge-seekers, other network dynamics are in play: if seeker 1 finds a satisfying answer to a hard question, he might tell seeker 2, saving her time and potentially error. Similarly, if groups can automatically compare patterns of hits and misses or vote on result quality, there are theoretical possibilities for benefit. To date, however, two main dynamics prevail: people query their network with natural language requests, or else they search different things (and in different ways) than do their friends. This long-tail characteristic of mass-market search engines thus far has limited the usefulness of social search.

New Advertising Models

While Lycos, AltaVista, and Google all began as research projects, the current state of search dictates that revenue models will be a consideration, and that normally implies advertising. Accordingly, personalization of search results is one artifact of GPS-enabled mobile search. If a user types "pizza" and is located at a particular latitude and longitude in Tampa, results from New York or Seattle can be ruled out with almost complete certainty. Other forms of personalization, often based on search history, are location-independent. Going forward, aggregation of behaviors across channels may become feasible: if a user's search engine knows that she looked for a plumber from her desktop machine at work, it might feed her appropriate results when she types in "hardware store" into her mobile device.

5.4.2 Issues and Concerns

Along with all of the power that search brings to advertisers, information producers, and searchers, many complex issues are emerging with regard to liability, privacy, and property rights.

What are the limits to search? For all the documents that crawlers can access, as many as 100 times more exist out of reach in databases or in sites excluded from crawls. More fundamentally, one cannot search unless one can name the object of the search. Machine parts are an obvious example: a gear came off a 1966 Case tractor and clearly is cracked, but search engines are of little use in finding replacements without the manufacturer's proprietary part number, which itself can change, be superseded, or disappear. In addition, new forms of digital literacy, and illiteracy, are emerging. Both the tools and their users are evolving: recall the shift away from Boolean queries, for example, and the limited success of natural language queries to date. The final limit to search is of course that nondigital items must somehow be brought into that realm.

What can and should search do with regard to privacy? Stalking or just curiosity browsing, a perennial issue at taxation and other authorities, is easy, and more records come online every year. In addition, when "digits never die," what are the paths for recourse when incorrect, outdated, or mishandled information falls into the path of a crawler? Similarly, as geolocation gets more powerful every year (with the rise of GPS-enabled cell phones, for example), the privacy implications of tools such as Google Street View and readily accessible overhead imagery remain unsettled.

Search connects to identity, both that of the searcher and that of the searched. When AOL released anonymized search histories for research purposes, two reporters (not algorithmic experts) reverse-identified a searcher almost immediately [24]. From advertisers' and investigators' perspectives, identity is connected to the problematic notion of search as a "database of intentions" [25, 26]. As the AOL data itself reveals, most of what is typed into a search bar is not necessarily a search at all or the least bit related to intention. Sex, obviously, is related to a substantial proportion of searches. Many are navigation: typing such terms as "gap.com" into Google rather than the address bar has become a habit for many. As the *New York Times* reported with regard to the woman their reporters reverse identified, "Explaining her queries about nicotine, for example, she said: 'I have a friend who needs to quit smoking, and I want to help her do it' " [24]. Simplistic assumptions about the relation of search terms to attention rarely shed light on complex issues of motivation, although Neil Entwistle was found guilty of murdering his wife outside Boston in 2006, in part because police recovered searches including "kill with a knife" and "quick suicide methods" and the prosecution introduced them as evidence at the trial [27].

If people's searches are connected to their identity, albeit in complex ways, who owns that data? The default setting for many Google products sends search histories to Google's headquarters, so the question becomes if an individual's search history will become like a medical record, owned by the outside party rather than the person who generated it. (This is certainly the case at Amazon: a customer cannot request his or her search or browsing history even if it is collected in the process of doing business.) Will there be secondary markets for search histories and if so, how will they operate? Already Google uses aggregated search data to identify both cultural trends and potential influenza outbreaks [28].

What does global search do to concepts of public records based in a locality? Saying that divorce filings or house sales should be public has a particular set of implications in a paper-based environment yet a completely different one in virtual space. Aerial photos that show the roofs of public facilities (such as schools or prisons) or attractive nuisances (suburban backyard swimming pools) or managed semi-public resources (trout streams, perhaps) each suggest some set of limits on infinite access.

Finally, a major question for search going forward concerns following the money: How does the arms race between search engines questing for better results and the quiet but highly influential Search Engine Optimization (SEO) industry shape the organization of, and access to, knowledge? Before Google was founded, the two inventors wrote the following in a scholarly journal [29]:

> [From] historical experience with other media, we expect that advertising-funded search engines will be inherently biased towards the advertisers and away from the needs of the consumers. Since it is very difficult even for experts to evaluate search engines, search engine bias is particularly insidious [20].

If search becomes the default tool for organizing much of the world's information, but the embedded assumptions remain invisible, yet commercially driven, what the world knows will be subtly shaped as well.

5.5 SEARCH AND HUMAN-CENTERED FUSION

How does the evolution of connectivity and search capabilities affect human-centered fusion? While not exhaustive, there are five initial areas in which the evolution of the Internet and Web will affect information fusion.

1. *Humans as soft sensors:* The very existence of the Internet, the blurring of communications and computing platforms, and the proliferation of mobile phone/computer devices lead to an increasing source of data based on human observations and humans acting as sensor platforms. Observational "boots on the ground" via both professionals and civilians leads to a huge resource of data and interpreted data related to observing and monitoring a situation. Thus, while increasing hard sensing capabilities will be readily accessible (see the next comment), the ability of in-place observers could lead to a shift from "trusting sensors" to "trusting human observers" at the location of interest. This trend will lead to new concepts in which

information fusion systems seek to task the observational crowd versus tasking traditional hard sensing resources.

2. *Ubiquitous sensing:* Internet connectivity and service-oriented architectures enable ad hoc sensor placement and use. Currently, services such as Twitpic (http://www.twitpic.com) allow the sharing of photos using the Twitter Internet service. Almost any sensing device can now be deployed and connected to the Web using Twitpic and cell phone devices. Hence, one can easily deploy ad hoc sensors throughout an environment. Sensing devices will proliferate to achieve a situation in which current inanimate objects such as buildings, roads, bridges, and automobiles will become self-aware (namely, monitoring their own state, health, and utilization), but capable of broadcasting that information throughout the Web. The data input to information fusion systems will no longer rely primarily on user-deployed sensor collection systems, but rather take advantage of ad hoc "other people's sensor" systems.

3. *Unintended sensor utilization:* With the items above, it is anticipated that human-centered information fusion systems will use existing sensors in unintended ways (namely, use them for purposes other than their original design). An example of the unintended use of sensor information is to observe traffic conditions by using cell phone traffic (namely, when an accident or poor road conditions adversely affect traffic, drivers call home to alert a spouse of friend that they will be late). Other examples include observing the distribution of garbage and discarded beer bottles after a football game to infer the population age distribution and the use of toll booth sensors to determine automobile occupancy (e.g., to enforce restrictions on high occupancy lanes).

4. *Information forensics:* Whenever a major event such as the Mumbai terrorist event occurs, the Web becomes a rich source of information to conduct forensic analysis. Key questions include: What was the timeline of the event or activity? Who were the communicating (e.g., Twittering and blogging) participants? What information was available via newscasts and open source reporting (and when)? What were the types of misunderstandings that occurred (by the perpetrators, defenders, or general public)? How did the participants use telecommunications and Web technologies (e.g., use of Twitter, Facebook, e-mail); and similar questions. Information forensics may involve searches to understand patterns, identification of precursors to an event or activity, identification of related activities or events, and general data mining activities to seek possible observables or correlated parameters. Searching the Web may become a routine form of analysis, analogous to monitoring the physical terrain for changes such as new structures. Similarly, one may monitor the Web for changes in the information landscape.

5. *Contextual data/model resource:* A major resource available on the Web to support human-centered fusion is contextual data for modeling and hypothesis evaluation: online weather data, access to webcams or distributed sensors, analysis of newsfeeds and blogs, access to Flickr and Google Earth to obtain photos of a localized environment, and access to models for terrain, communications, observability, social network activity, and more. In this case, the Web data becomes a background that supports situation awareness, providing the background "information terrain" as the backdrop for evaluating a current situation, activity, or event, and providing a sanity check as new information is received. In addition, Web resources are becoming available to support remote virtual experimentation and modeling [6].

It is clear that the increasing information on the Web and new search methods will play an important role in changing the concept of information fusion to a new era of human-centered fusion. As open-source Web-based information is increasingly considered as a primary information source, new search engines, metadata generation techniques, and information services will appear. In the near future, it would not be unexpected to see a new Googledata or Mozilldata type of search engine that focuses especially on finding data (sensor data, models, archived images, signals, and analyzed vector data. For an example related to marine data sets, see http://seamap.env.duke.edu/help/datasets.html) for specified applications and areas of interest. Moreover, new commercial enterprises may develop to specialize in the creation, calibration, and assessment of Web data, along with services to support the data-finding and data-understanding.

References

[1] Palfrey, J., and U. Gasser, *Born Digital: Understanding the First Generation of Digital Natives*, New York: Basic Books, 2008.

[2] Sherman, C., and G. Price, *The Invisible Web: Uncovering Information Sources Search Engines Can't See*, Medford, NJ: CyberAge Books, 2001.

[3] http://www.zakon.org/robert/internet/timeline/.

[4] Witten, I. H., M. Gori, and T. Numerico, *Web Dragons: Inside the Myths of Search Engine Technology*, San Francisco, CA: Morgan Kaufmann Publishers, Elsevier, 2007.

[5] http://www.webreference.com/authoring/search_history/.

[6] Ko, C. C., and C. D. Cheng, "Web-Based Laboratories for Internet Remote Experimentation," in *World Review of Science, Technology and Sustainable Development*, Vol. 5, No. 3.4, 2008, pp. 301–314.

[7] Warsh, D., *Knowledge and the Wealth of Nations: A Story of Economic Discovery*, New York: Norton, 2006.

[8] Friedman, T., *The World Is Flat: A Brief History of the Twenty-First Century*, New York: Farrar, Strauss, and Giroux, 2005.

[9] Anderson, C., *The Long Tail: Why the Future of Business Is Selling Less of More*, New York: Hyperion, 2006.

[10] Taleb, N., *The Black Swan: The Impact of the Highly Improbable,* New York: Random House, 2007.

[11] In 1997, 1% of sites attracted 70% of traffic; 2003 figures reaffirmed the hypothesis with weblog data. Nielsen, J., "Zipf Curves and Website Popularity," http://www.useit.com/alertbox/zipf.html.

[12] Stockwell, F., *A History of Information Storage and Retrieval,* Jefferson, NC: McFarland, 2001, p. 35.

[13] Wright, A., "Exploring a 'Deep Web' That Google Can't Grasp," *New York Times*, February 22, 2009, http://www.nytimes.com/2009/02/23/technology/internet/23search.html?th&emc=th.

[14] Smith, G., *Tagging: People-Powered Metadata for the Social Web,* Berkeley, CA: New Riders, 2008.

[15] Dean, J., and S. Ghemawat, "MapReduce: Simplifed Data Processing on Large Clusters," *OSDI'04: Sixth Symposium on Operating System Design and Implementation*, San Francisco, CA, December 2004.

[16] O'Reilly, T., "The Architecture of Participation," June 2004, http://www.oreillynet.com/pub/a/oreilly/tim/articles/architecture_of_participation.html.

[17] Johnson, J. P., "Collaboration, Peer Review and Open Source Software," May 2004, http://ssrn.com/abstract=535022.

[18] Campbell, P., "Report of Nature's Peer Review Trial: Despite Enthusiasm for the Concept, Open Peer Review Was Not Widely Popular, Either Among Authors or by Scientists Invited to Comment," Nature.com, December 20, 2006, http://blogs.nature.com/peer-to-peer/2006/12/report_of_natures_peer_review.html.

[19] American Press Institute, "Paid Content: Newspaper Economic Action Plan," May 2009.

[20] Nahin, R. L., et al., "Costs of Complementary and Alternative Medicine (CAM) and Frequency of Visits to CAM Practitioners: United States, 2007," National Health Statistics Reports, Centers for Disease Control and Prevention, National Center for Health Statistics, July 30, 2009.

[21] Broom, A., "Medical Specialists' Accounts of the Impact of the Internet on the Doctor/Patient Relationship," *Health,* Vol. 9, No. 3, 2005, pp. 319–338.

[22] Wang, J. Z., J. Li, and G. Wiederhold, "SIMPLIcity: Semantics-Sensitive Integrated Matching for Picture Libraries," *IEEE Transactions on Pattern Analysis and Machine Intelligence*, Vol. 23, No. 9, 2001, pp. 947–963.

[23] Datta, R., et al., "Image Retrieval: Ideas, Influences, and Trends of the New Age," *ACM Computing Surveys*, Vol. 40, No. 2, article 5, April 2008, pp. 51–60.

[24] Barbaro, M., and T. Zeller Jr., "A Face Is Exposed for AOL Searcher No. 4417749," *New York Times*, August 9, 2006, http://mrl.nyu.edu/~dhowe/TrackMeNot/NYTimes_AOL_Exposed.htm.

[25] Battelle, J., *The Search: How Google and Its Rivals Rewrote the Rules of Business and Transformed Our Culture,* New York: Portfolio, 2005, pp. 1 ff.

[26] Leonhardt, D., "The Internet Knows What You'll Do Next," *New York Times*, July 5, 2006.

[27] Ellement, F. R., and D. Abel, "Entwistle Talked of Finding Bodies," *Boston Globe*, June 7, 2008.

[28] http://www.google.com/trends and http://www.google.org/flutrends/.

[29] Brin, S., and L. Page, "The Anatomy of a Large-Scale Hypertextual Web Search Engine," 1998, http://infolab.stanford.edu/~backrub/google.html.

Chapter 6

Data Visualization and Understanding

More than a quarter-century has passed since the U.S. publication of Edward Tufte's landmark book, *The Visual Display of Quantitative Information* and the English translation of Jacques Bertin's *Semilology of Graphics* [1, 2]. In that time, computer screens and other projection tools have emerged as a powerful medium challenging the primacy of paper, previously the default tool of choice for information visualization. The field is now exploiting new display technologies (such as flexible OLEDs [3]); new computing platforms, including iPhones and similar devices; and ever-increasing computing power such as can be found in a PlayStation 3 or other contemporary game platform. For visualization to capitalize on the power of these and other technologies, information architecture must increase in sophistication, usability, and explanatory leverage. This chapter begins with a brief overview of visualization, provides examples of successful execution, touches on recent advances, and concludes with a comparison of visualizations of physical versus nonphysical data before concluding with a discussion of lessons and opportunities.

6.1 INTRODUCTION

The need for effective information visualization results from both supply and demand. From the supply perspective, business applications at both the desktop and enterprise levels include tools for charting, graphing, and report generation. In the consumer realm, expectations are being raised by election night coverage, complex weather mapping, and even gaming. On the demand side, data is generated by more sources and available to more users every year. "Data glut" may be a cliché, but tools such as search engines can intensify it: Google recently announced that 1 trillion Web pages had been indexed. For context, if each of the 32 million books in the Library of Congress averages 300 pages, that is less than 10 billion physical pages, and these individual units reside in nothing resembling a unified, organized, searchable repository. As video moves to the Internet via YouTube and other mechanisms, data volumes grow at even faster rates, commensurately complicating the task of understanding.

6.1.1 Definitions

Definitions for information visualization have proliferated. For our purposes, Williams et al. [4] have provided a useful starting point: "a cognitive process performed by humans in forming a mental image of a domain space. In computer and information science it is, more specifically, the visual representation of a domain space using graphics, images, animated sequences, and sound augmentation to present the data, structure, and dynamic behavior of large, complex data sets that represent systems, events, processes, objects, and concepts." More recently, the movement from paper to screen-based computer graphics has led to a new definition of visualization: "the use of computer-supported, interactive visual representations of data to amplify cognition" [5].

Such comprehensive statements raise many issues, some of which we will discuss further. Foremost among these is the question of hardware versus "wetware": How is the boundary between information presentation through technology and human cognition understood and managed in a given instance? That is, how does what is projected relate to what is understood? Subsidiary questions revolve around the oft-stated but poorly defined notion of "insight." How well does a visualization answer precise versus vague questions on the part of a given viewer? Indeed, how well does a visualization raise new questions as opposed to answering existing ones? Finally, visualization can address the essential human characteristic of curiosity [6], a trait that, much like pornography in the view of the U.S. Supreme Court, we know when we see it, making assessment difficult, if not impossible.

6.1.2 Objectives of Information Visualization

Computer-driven information visualization can be understood within a long history of paper-based efforts to help people understand complex phenomena. From early tabular and graphic representations of temperature and precipitation until 1982—when *USA Today* made a full-page set of color maps and tables a trademark of the upstart publication—and beyond, tools for understanding weather and climate have

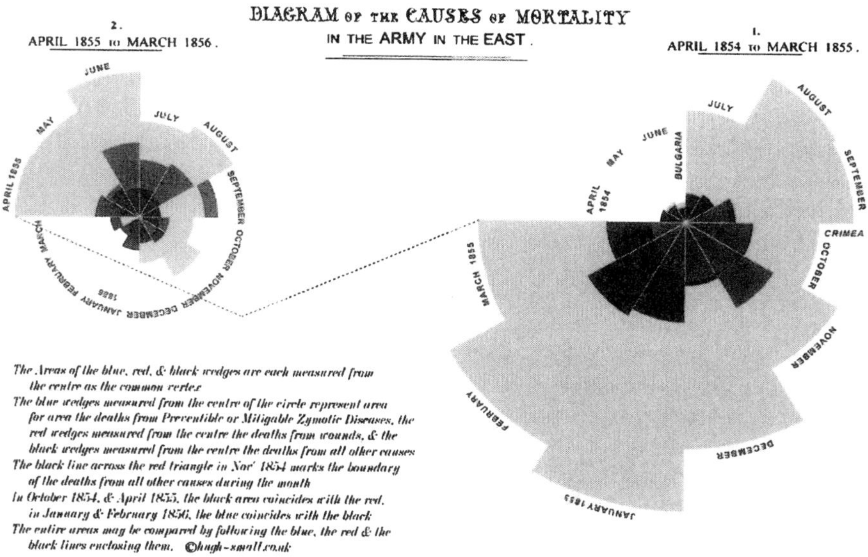

Figure 6.1 Polar area map of sanitation-related deaths among British soldiers in the Crimean War.

helped lead the state of information visualization. More recently, graphics workstations were overrepresented in television studios as an arms race of weather-casting helped advance the state of the field. The weatherpeople can boast results: many more people can understand a Doppler radar image than can grasp binomial distributions, bid-ask spreads, or tree maps.

Historically, many visualizations have resulted from individuals who wanted to change their world: Florence Nightingale's striking graphics (she refined the polar area map; see Figure 6.1) relating conditions in military hospitals after the Crimean War helped persuade Queen Victoria to initiate broad reforms in public sanitation.

More recently, the tools built by Hans Rosling and his colleagues at Gapminder are clearly aimed at increasing public awareness of public health, developmental economics, and other social issues (Figure 6.2). That figure shows the rate of infant mortality (before age 5) to normalized per capita income for various countries (the size of the bubbles indicate the size of the country's population).

Other types of Gapminder graphics show relationships such as life expectancy at birth versus average income by country (http://graphs.gapminder.org/world/).

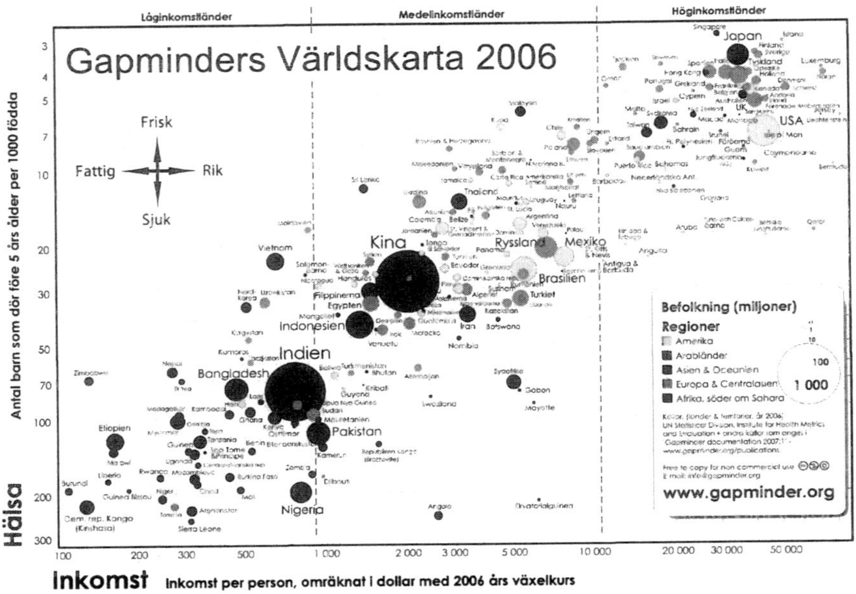

Figure 6.2 Gapminder world chart (original Swedish version) mapping GDP per capita and mortality before age 5.

6.1.3 A Brief Tour of Information Visualization

Given that information visualization exists in (and between) both information systems and human brains and currently consists of more craft than standards and more art than science, generalizing is difficult. Implicitly or explicitly, all visualizations answer roughly the same set of questions:

- How are similarity and difference conveyed?
- Is time static (as in a pie chart) or dynamic, as in many line graphs or slider-bar tools?
- How much granularity is sacrificed for "glanceability," and how much comprehension time and difficulty are required to deliver details?
- Is causality intended to be conveyed? Can it be unreasonably inferred?
- How does space function in the representation? Do proportions relate to some ground truth?

- What do the colors convey? Are colors used in standard (green = proceed safely; red = danger) or nonstandard ways?
- How are the reliability, timeliness, accuracy, precision, and other attributes of the underlying data represented in the visualization [7]?
- Unlike paper, computer visualizations can be interactive. How easily can the user learn to reposition, zoom in, reset a baseline, and otherwise get the visualization to respond to his or her actions?
- How likely is ambiguity? Why might the display be read three different ways by three different people?
- Independent of the graphical tool(s) chosen, does the visualization ask appropriate questions of the data?

6.2 UNDERSTANDING VISUALIZATIONS

The state of information visualization is still relatively immature insofar as few standards or objective criteria exist. Any given tool can be placed in numerous frameworks and can be judged by multiple yardsticks. Some basic questions can be asked of any visualization, however, relating to the data being visualized, the tools of representation, and the users to whom the visualization is addressed.

6.2.1 Taxonomies

Given the wide swath covered in both concept and practice by the overlapping notions of information visualization, data visualization, scientific visualization, geographic visualization, statistical graphics, and other fields, delimiting the various fields can be difficult. Are models grouped by visual technique, by the nature of the underlying data, by the professional domain (e.g., medicine versus aeronautical engineering), or in some other fashion? To what extent are characterizations based on "push" (technology) and "pull" (cognitive processes and outcomes)? How does the problem under consideration shape the visualization, independent of the nature of the data being represented? While such questions are far from being resolved, three efforts deserve recognition.

Tory and Möller [8] based their taxonomy on characterizations of models of the data rather than on characteristics of the data itself. They proposed dividing design models into continuous and discrete categories: to oversimplify, continuous models allow interpolation, while in discrete models (male versus female, for example), interpolation is not meaningful. Turning to the display attributes such as color, spatialization, and transparency, models fall along a continuum. At one extreme, attributes are given, as in latitude/longitude or molecular structures. At the other extreme, the researcher can choose the aspects of spatialization (as in file structures). In the middle of the continuum, the problem is partially constrained. The resulting taxonomy is shown in Table 6.1.

From the viewpoint of a consultant rather than an academic, Robert L. Harris [9] developed an "illustrated reference" of visualization techniques. The book is organized alphabetically, from "abscissa" to "zigzag graph," making it problematic to find a visual technique for which one lacks a name. For a practitioner, Harris provides valuable details such as axes and legends, and for a user of visualizations, he identifies common modes of deception or ambiguity such as broken scales and pseudo-3-D effects.

Table 6.1

High-Level Visualization Taxonomy from Tory and Moller [8], Constraints Relating to Spatialization

	Display Attributes		
	Given	*Constrained*	*Chosen*
Continuous	Images (e.g., medical)	Distortions of given /continuous ideas (e.g., flattened medical structures, 2-D geographic maps, fish-eye lens views)	Continuous (high-dimensional) mathematical functions
	Fluid/gas flow, pressure distributions	Arrangement of numeric variable values	Continuous time-varying data, when time is mapped to a spatial dimension
	Molecular structures (distributions of mass, charge)		Regression analyses
	Globe—distribution data (e.g., elevation levels)		
Discrete	Classified data/ images (e.g., segmented medical images)	Distortions of given/discrete ideas (e.g., 2D geographic maps, fish-eye lens views)	Discrete time-varying data, when time is mapped to a spatial dimension
	Air traffic positions	Arrangement of ordinal or numeric variable values	Arbitrary entity-relationship data (e.g., file structures)
	Molecular structures (exact positions of components)		Arbitrary multidimensional data (e.g., employment statistics)
	Globe—discrete entity data (e.g., city locations)		

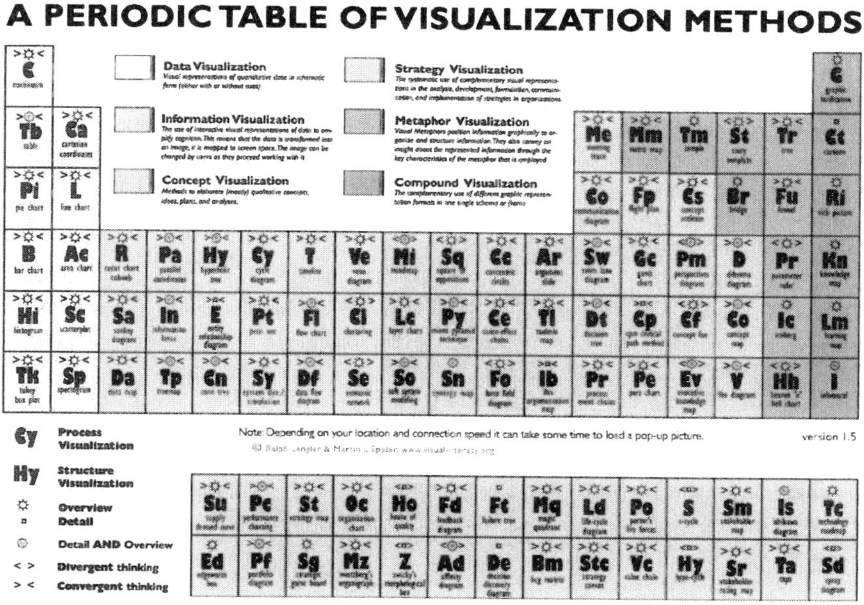

Figure 6.3 Lengler and Eppler's periodic table of visualization methods.

Two Swiss professors, Ralph Lengler and Martin Eppler, created a visual tool for understanding visualization tools [10]. Basing their model on what many consider the most powerful visualization tool in history, the periodic table of the elements, Lengler and Eppler created a grid for 100 different visualization methods, placing like near like, distinguishing between visualizations of process and structure, and attempting to capture other distinctions. A mouse rollover delivers an example of the given technique (Figure 6.3).

6.2.2 Assessment and Evaluation

Information visualizations can be assessed along multiple dimensions, but none of these can claim ultimate authority. Because of the nature of academic publishing, empirical studies cannot focus on broad questions whose results are inconclusive and may takes weeks or months to appear. The following list, which draws heavily on Carpenter [11], illustrates the difficulty of assessing any given visualization in a standardized yet context-sensitive manner. At base, most studies of visualization focus more on human interaction with hardware—input—than on human cognitive and behavioral consequences.

Human computer interaction (HCI) investigators will focus on tasks: when asked, how readily do research subjects successfully zoom, filter, or drill into the details of the data? One major difficulty in HCI research is identifying the right experimental subjects: too little familiarity with an application domain or toolset can be as much of a bias as too much expertise when selecting subjects for a statistically significant subject pool. All subjects will implicitly or explicitly benchmark a lab experience against prior experience, whether a commercial tool, Google, or a video game. Similarly, usability is often understood (and represented) in relative rather than absolute terms.

Coming from the perspective of perceptual psychology, researchers might pay attention to such essential (but incomplete) factors as readability. Because cognitive reasoning tasks, especially for complex visualizations, are often only roughly defined, asking subjects to compare, associate, or categorize data can lead to unsatisfactory results. When it comes to higher-order tasks such as spotting trends or causal relationships, individuals will vary in their approach to an experience with a given visualization; furthermore, the cognitive lightbulb may illuminate days or weeks after the experiment.

A question that crosses disciplines underlies the entire field is: How well does a given visualization provide insight into the data? Unlike task completion, discovery is difficult to script. As one researcher notes, this aspect of the field constitutes "answering questions you didn't know you had" [12]. When one encounters these moments, direct causation is typically elusive. Furthermore, individuals in a lab behave very differently from members of a team, which is frequently the locus of a search for understanding based on large quantities of data. Other dynamics of real-life visualization—as in first-responder dispatch, combat scenarios, or other life-critical situations—are extremely difficult to model using subjects most commonly available to university researchers.

Finally, even though the academic literature on cognitive heuristics and biases is well founded [13] and recognized by a 2002 Nobel Prize in economics, few visualizations are evaluated on the basis of the decisions they support: a given screen may be legible, prompt reliable task completion, and look pretty, but until it can drive effective decisions, the other success criteria would seem to be secondary considerations. The fields of visualization and decision support would appear to have much to teach each other.

Given these shortcomings of the current state of the discipline, at least one researcher [14] has suggested merging aesthetic judgment and technology assessment. Attempting to guide the discussion "away from implementation details and single mouse clicks to the meaning of a visualization," the concept of visualization criticism draws heavily on the tradition of design critiques which, while subjective, are based on a long and deep body of practice.

6.2.3 Visual Analytics

An emerging area is the concept of visual analytics. This is the concept of integrating analytical reasoning techniques with highly interactive visual interfaces. It extends traditional studies of geographical information systems into general issues of data analysis, data mining, data visualization, and interactive (human/computer) analysis using visualization tools. An overview of the research and development agenda for visual analytics is provided by the Web site for the National Visualization and Analytics Center (http://nvac.pnl.go/agenda.stm). The site provides a link to download a book entitled *Illuminating the Path: The Research and Development Agenda for Visual Analytics*, which identifies grand challenges, the concept of analytical reasoning, visual representations and interaction techniques, data representations and transformations, and other concepts.

Five regional visualization and analytics centers (RVACs) have been established in the United States: (1) at The Pennsylvania State University, (2) at Purdue University and Indiana University School of Medicine, (3) at Stanford University, (4) at the University of North Carolina at Charlotte and Georgia Institute of Technology, and (5) at the University of Washington. The RVAC at Penn State is coordinated through the GeoVISTA Center hosted by the Department of Geography (http://www.geovista.psu.edu). The GeoVISTA Center provides access to an open source toolkit called the GeoViz toolkit (http://www.geovista.psu.edu/geoviztoolit.index.html), which is available for supporting geographical visual displays, analysis of large data sets, and development of spatio-statistical views of data.

6.3 SUCCESS STORIES

Successful information visualizations can be found in several domains:

- Hollywood and the gaming market are leading a charge toward effective, practical, and multidimensional visual tools. A quick look at any SIGGRAPH conference program illustrates the overlap as featured speakers come from both computer science centers like Carnegie Mellon and animation houses such as Pixar.
- 3-D has become a day-to-day reality in architecture, manufacturing, and, again, gaming, and this trickles down to consumer applications such as kitchen design tools. Even state-of-the-art roller coasters are being rendered in CAD on YouTube for mass enjoyment.

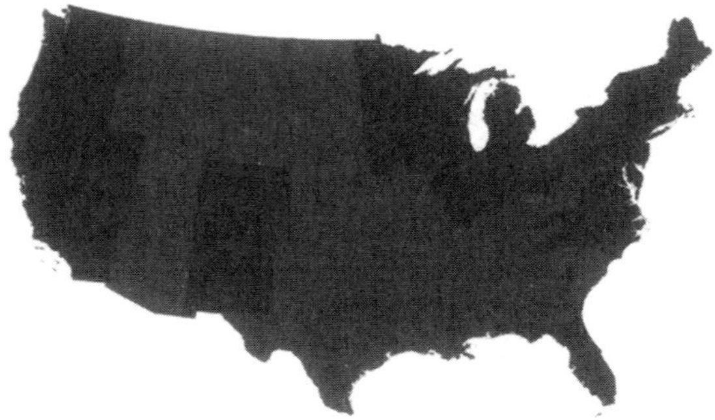

Figure 6.4 The 2008 electoral results in standard format.

- Election-night returns have traditionally featured innovations in data representation. The red state/blue state distinction has entered common spoken usage, for example. But maps as a data template lack the ability to correlate population with area, so Rhode Island, with 1.1 million citizens, is twice as populous, but only 1/60th of the land area of Wyoming. Mark Newman, a University of Michigan physics professor, has corrected this problem with cartograms, which rescale states by population. The 2008 electoral results look familiar in standard mapping in Figure 6.4, which simply shows a traditional U.S. map with states colored in red (for a state whose voters favored the Republican party) or blue (for a state whose voters favored the Democratic party). Figure 6.5 shows a revised view in which the size of the states is varied to reflect the number of votes. With the population correction [15], the Plains states and mountain west recede in impact (Figure 6.5). (In Figures 6.4 and 6.5, the traditional color red is show as light gray, and the color blue is shown as dark gray.)

Figure 6.5 The 2008 popular vote results mapped as a cartogram.

- Since the release of its Flex version 2 product in 2006, which extended Flash functionality to more programming environments, Adobe has supported the rapid development (in both senses of the term) of visually attractive, data-driven visualizations: the Flex application showcase at flex.org features a wide variety of database-driven shopping, monitoring, configuration, and wayfinding tools. Some are extremely handsome and useful, and even some of the visually "flat" examples possess a high information density.

- Mapping remains important. Through the release of APIs from the likes of the UK Ordnance Survey, ESRI, Microsoft, and Google, developers can build data-rich, geographically useful tools more easily than ever before. Once again, Flex can accelerate the process.

- Transparency in real or implied 3-D data volumes can allow exceptions to stand out more clearly. In 2-D, transparency preserves the baseline information.

- Time can be manipulated via sliders and other intuitive tools, effectively creating animations. True data richness, as in this example from the Boston Federal Reserve Bank (Figure 6.6), is well served by easy comparison across time and county, instantly obvious navigation, subtle but clear use of color for information, and appropriate scale: doing such a tool on a national level would be unfeasible and lose appropriate granularity.

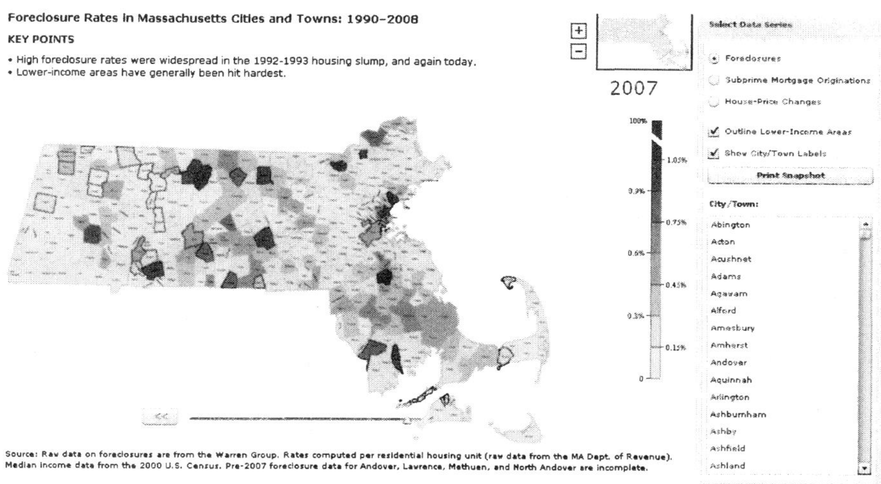

Figure 6.6 Federal Reserve Bank of Boston interactive map of subprime loan originations and foreclosures in Massachusetts 1990–2007 (http://ww.bos.frb.org/economic/dynamicdata/module1/bmap.html#).

There is no shortage of activity in the field, samples of which can be experienced at Flex.org, Visual Complexity, or IBM's Manyeyes (http://manyeyes.alphaworks.ibm.com/manyeyes/). Some work is truly stunning, and global centers of design leadership are emerging. Even so, the fundamental tension quickly becomes evident: words like "galleries" suggest that we are viewing works of art, and in many instances the work should be in museums. But art by definition is unique; visualization has yet to be brought to the masses of managers, citizens, and students who have something to say but lack the tools, grammar, and training to create the beautiful. In short, the task of helping high levels of information visualization migrate from the artist to the worker remains unaccomplished.

6.4 COMMERCIAL TOOLS

Today's information worker has a wide variety of information visualization tools from which to choose. Arranged roughly from easiest and least expensive to most difficult and costly, these tools fall into the following broad categories.

6.4.1 Desktop Office Suite

For all the investment in enterprise packages in both public and private sectors, spreadsheets and presentation graphics generate vast numbers of visualizations. The current state of practice within Microsoft Office delivers hundreds of standard templates for flow charts, graphs, pivot tables, and other representations. Cost and difficulty are generally low. Figure 6.7 shows a spreadsheet with the ability to quickly select graphical representations of the data by simply selecting from a menu of techniques such as bar charts, scatter plots, and pie charts.

6.4.2 Scientific and Technical Packages

Numerous domain-specific tools apply to specialized markets: MATLAB, Mathematica, Chemdraw, GenePalette, Revit, Maya, and many others. These tools are extremely powerful, capable of generating 3-D effects in both still and animated output. The lines between static design and dynamic simulation, between architecture and entertainment, and between physical reality and silicon representation blur more each year. Output from a given tool may be used in a video game, a construction document, or a marketing focus group, for example.

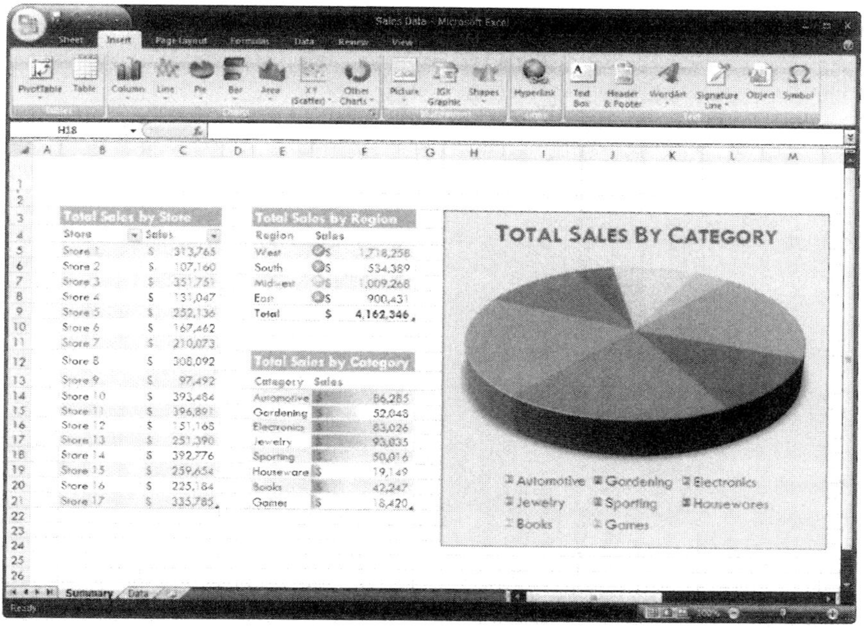

Figure 6.7 Charting capabilities within Microsoft Excel.

As an example, Paul Nylander graphed the Magnus effect on a spinning baseball in Mathematica. Figure 6.8 dramatically shows the phenomenon in which a spinning ball in a fluid creates a whirlpool of fluid around itself, which results in differential pressure effects.

6.4.3 Statistical Packages

Several statistical packages allow general-purpose charting, graphing, and visual analysis. Among these are products are SAS, SPSS, and Stata. Many of these packages are also positioned in the enterprise data analysis market, known as business intelligence (BI). Statistical packages generally require specialized training and include proprietary scripting languages. An example of a 3-D

Figure 6.8 Streamlines around a spinning baseball visualized in Mathematica.

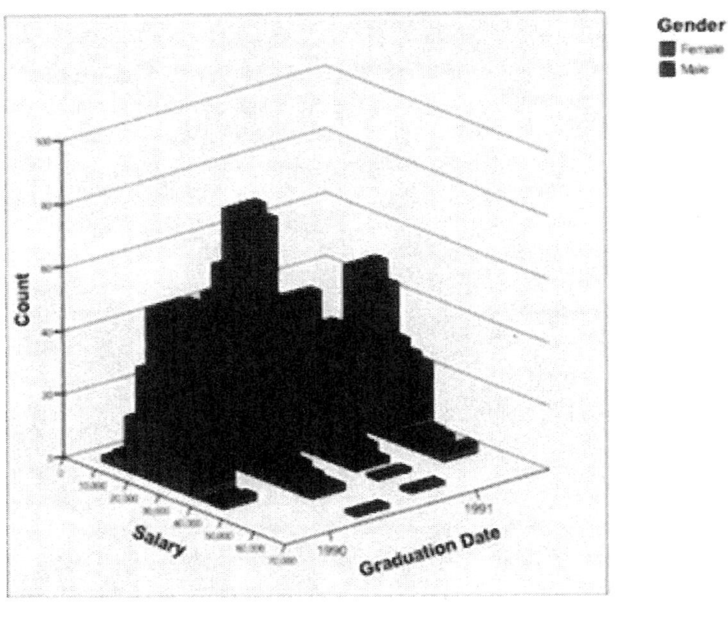

Figure 6.9 Complex visual output from the SPSS statistical package.

representation of data is shown in Figure 6.9, where in that figure, a 3-D histogram shows the differences in salaries obtained by females versus males as a function of graduation date.

6.4.4 Geographic Information Systems

At the same time that availability of GPS data has driven the rise of an entire subsection of the consumer electronics industry, tools for geographic analysis are also advancing rapidly. Commercial vendors such as ESRI (Figure 6.10) have been joined by Google, whose mapping and geographic imagery APIs have driven a generation of experiments in academic settings, and mashups (lightweight combinations of feeds from multiple sources, for example, a ski-lift camera, a weather report, and a map) in the general population (Figure 6.11). Thus, numerous overlays can be developed to highlight features such as terrain, political boundaries, businesses, and transportation nets.

6.4.5 Business Intelligence Tools

As part of an ongoing trend toward consolidation of the enterprise application software market, visualization vendors have been incorporated into broader package or service offerings. In 2007 alone, Oracle bought Hyperion, IBM acquired Cognos, and SAP bought Business Objects. Combined with a strong offering from Microsoft, these players dominate the market, leaving only SAS and

Figure 6.10 Classic GIS application: electrical infrastructure in ESRI ArcMap.

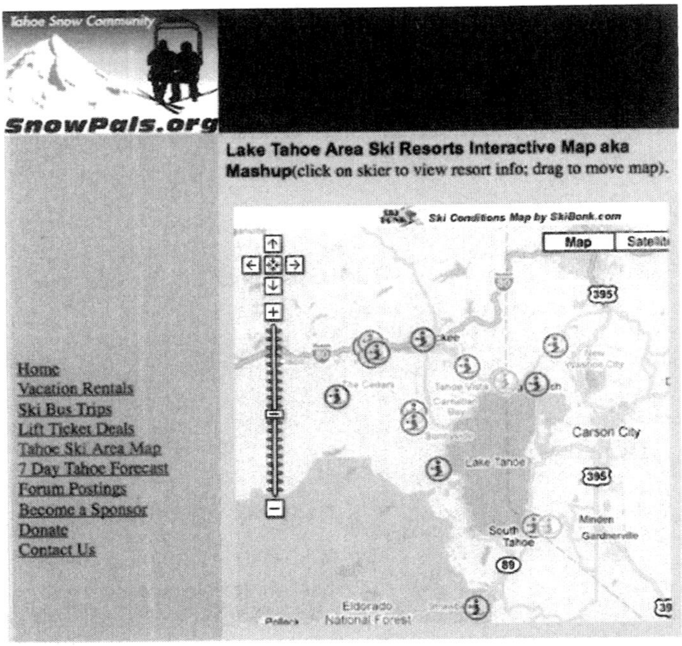

Figure 6.11 Mashup utilizing Google maps, trail condition information, and weather forecasts.

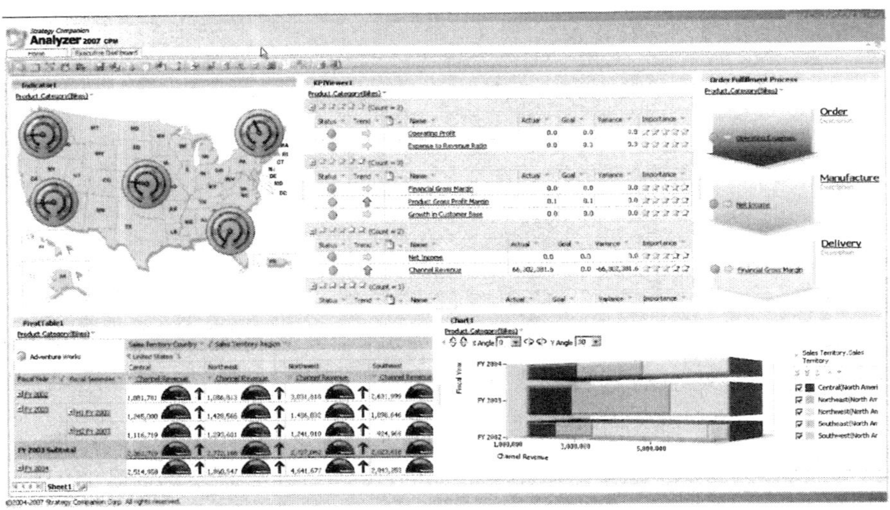

Figure 6.12 Hypothetical executive dashboard.

Microstrategy as leading standalone contenders. Business intelligence software, while not particularly expensive or difficult to use, does require sometimes extensive work on the data to be analyzed. Between standardization on format and nomenclature, explication of assumptions across operating units, and shared storage and access at the enterprise level, the data warehousing and data mining components of a BI effort can run into millions of dollars before a single report is generated. While executive dashboards (Figure 6.12) are a common output from a BI system, other visualizations such as basic tables, charts, and graphs figure prominently (Figure 6.13). Figure 6.12 shows a conceptual dashboard that seeks to provide a manager or executive with information about sales, order deliveries, product categories, and income by geographical location. By contrast, Figure 6.13 is an example of more traditional graphics used in business analysis. Table 6.2 provides examples of types of visualization tools for different application domains.

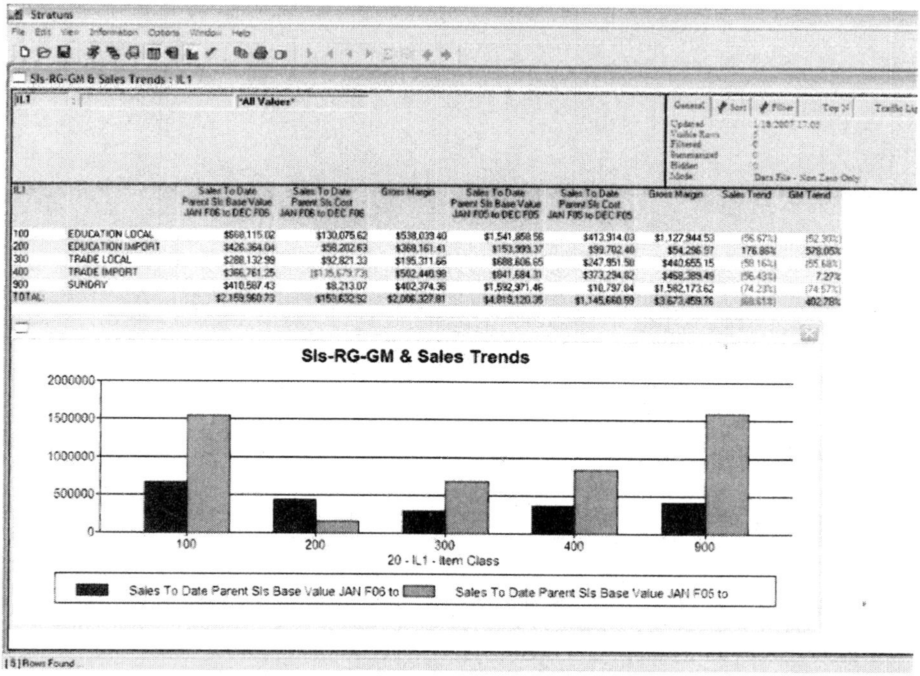

Figure 6.13 Enterprise data visualization in a business intelligence application.

Table 6.2

Examples of Visualization Tools for Selected Applications Domains

Visualization Tool Category	*Example*	*Ease of Use*	*Cost*	*Notes*
Desktop office suite	Excel	High to medium	Low	The default in many organizations
Scientific and technical packages	ChemDraw	Moderate for domain experts	Medium	Tools training now a significant percentage of many professions (e.g., architecture)
Statistical packages	SPSS	Difficult	Medium	
Geographic information systems	Google Earth/Maps, ESRI ArcMap	Variable	Low (free) to high	Move toward standard APIs contrasts with proprietary scripting languages in statistics packages
Business Intelligence tools	Business Objects, Oracle	Variable, depending on user's role	Dependent on associated changes to process, data, infrastructure	

6.5 VISUALIZING PHYSICAL DATA

An old saying asserts that "the map is not the terrain." Visualizations of physical phenomena are by definition abstractions, which means that interpretation, selection, and representation issues confront both makers and viewers of these visualizations. Compared to nonphysical phenomena, concepts, and data, which we will discuss presently, physical data is more straightforward, but significant considerations still inform the craft of visualization in the physical domain.

A tangible example is provided by Harry Beck's classical visualization of the London Underground first completed in 1931 and revised thereafter [16]. Compared to the predecessor map, which is shown in Figure 6.14, Beck's elegant abstraction (Figure 6.15) was much more readable. It does suffer, however, from factual errors: stations on different lines that are only a few hundred meters apart are shown as far away, leading riders to make two transfers and walk down long tunnels when, had they been aboveground, they could have seen how close the stations were in reality.

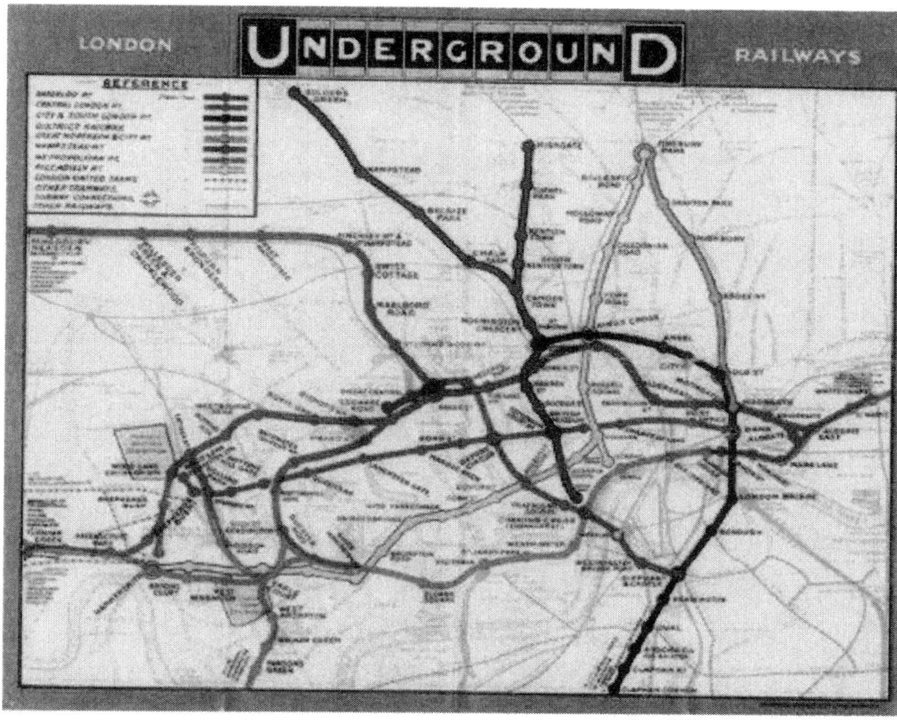

Figure 6.14 London Underground map dating from 1930.

Beck, an electrical draftsman by training, introduced the convention of a grid, allowing only 90 and 45 degree angles, much like in circuit diagrams. This artifice largely eliminated topography from the design brief, concentrating instead on the relative location along a given line: distances between stations on different lines are often vastly out of proportion. Other key elements of the map's success—use of white space, color-based simplification, and modern design cues including typography and limited symbolism—came from Frank Pick, the design-minded head of publicity for London Underground. Weighing the importance of usability for the task at hand against the physical reality being represented is a key step, and the importance of Beck's contribution, as well as its limitations, is reflected in London Transport's decision to call the aid not a map but a Journey Planner.

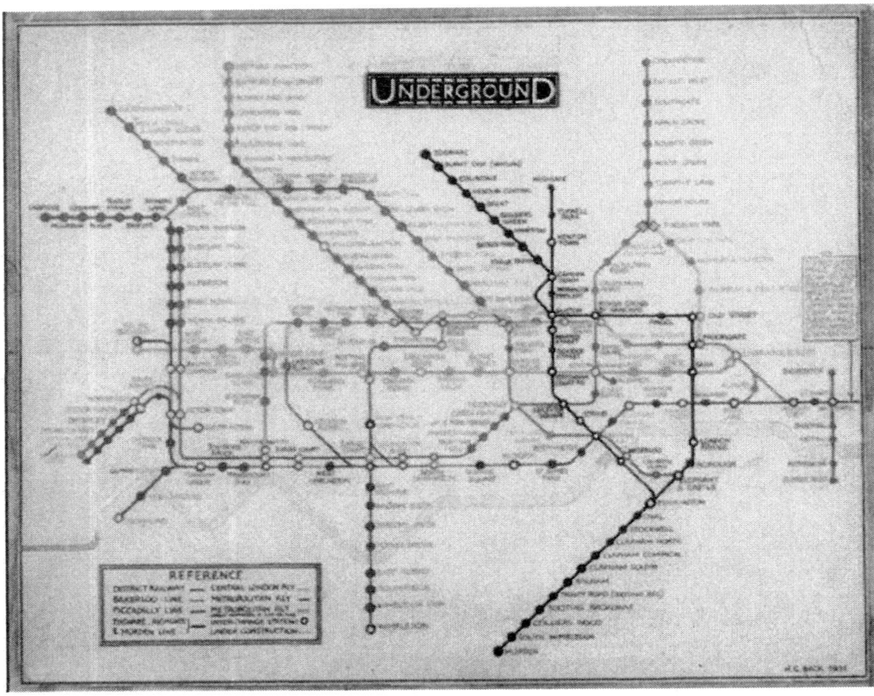

Figure 6.15 Early version of iconographic London Underground map [17].

6.6 VISUALIZING NONSPATIAL DATA

The task of nonspatial data visualization is particularly difficult because good displays must create spatial representations of nonspatial data. While the concept of geographic information displays is relatively straightforward (using one or more variations of a map and presenting overlaid information on the map), the representation of nonspatial data can be more challenging. This is not new: linear representations have conveyed time for millennia, and pie charts have become handy shorthand for subsets of a whole. Good maps remain the gold standard, but enjoy the advantage of being a spatial representation of space rather than something less tangible. Consult a U.K. Ordnance Survey map, or a fine nineteenth-century sample from any number of countries, and compare the quality to the nonspatial representations we encounter every day: *USA Today* visuals, executive dashboards, or owner's manuals. In most cases, the antique remains superior to the modern.

A powerful visualization known as a tree map has proven very useful for nonphysical data. *Smart Money*'s The Map of the Market (http://www.smartmoney.com/map-of-the-market/), which visualizes daily stock-market performance, is probably the best known tree map, a generic example of which is shown in Figure 6.16. Information domains are formed of rectangles, each of which includes component entities, sized proportionately to population, market capitalization, risk, or other variables and available for inspection by mouse rollover. The visualization provides at-a-glance awareness of the state of the entire domain, a given sector, or individual components.

Going forward, information architects are challenged to create readable, repeatable conventions for such abstractions as risk, intellectual property (patents are a poor proxy for human capital, for example), and attitudinal information such as customer satisfaction or confidence in government. Search and visualization have much to offer each other: semi-arbitrary lists of text-string matches remain hard to make visual (concepts are notoriously difficult to map spatially, in contrast to the elegance of the periodic table of the elements, to take a classic example). Current social network maps, especially those of large social graphs such as Facebook, quickly grow useless, as Figure 6.17 illustrates. Attempting to show a

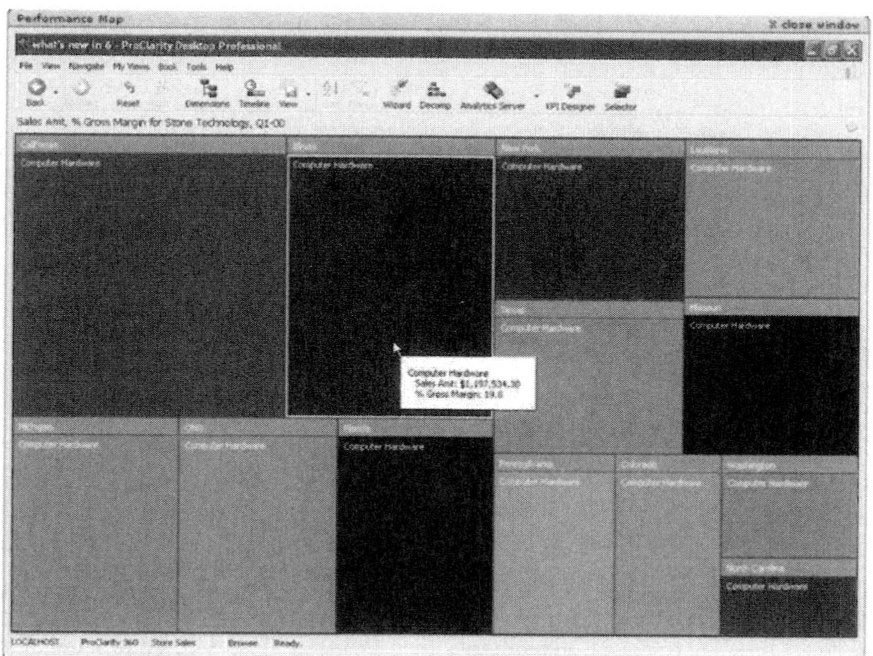

Figure 6.16 A generic tree map, or heat map, in which the rectangles' size is proportional to market size and color indicates growth rate.

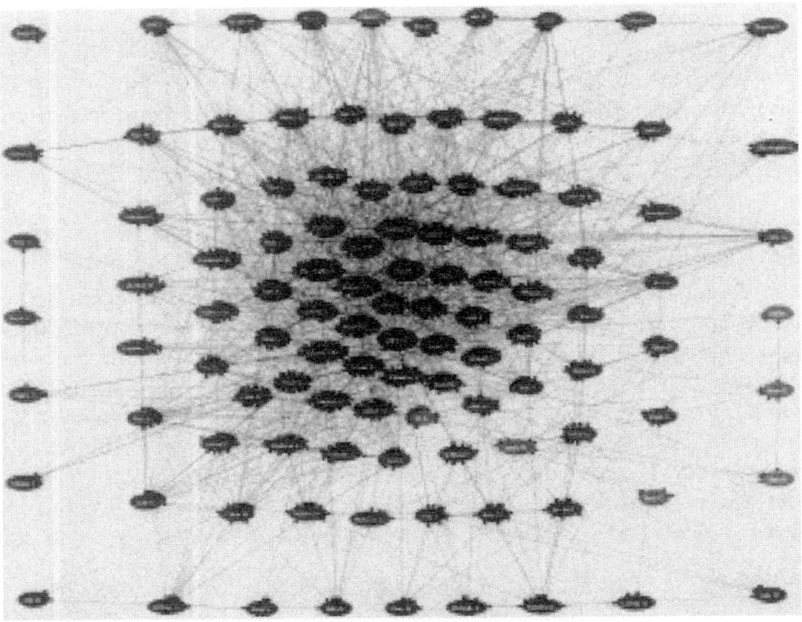

Figure 6.17 A graph of social network connections cannot scale past a tiny number of users.

dynamic social network using a graph (e.g., showing individuals as nodes in a two-dimensional graph and social interconnections as links between nodes) is a common representation but may not be satisfying because of limitations in showing the character of the social links (e.g., type of connection, strength, duration, interactions among multiple people) and other factors.

6.7 LESSONS AND OPPORTUNITIES

Some precedents may be useful. The history of sailing and shipping is rich with examples of various parties agreeing on conventions (port and starboard do not vary in different countries the way rules for automobiles do) and solving problems of conveying information. Container ships interlock regardless of carrier while being handled at countless global ports [18]. The Beaufort wind scale arose from the need for agreed-upon metrics for measuring wind aboard a ship, a matter of great practical importance. Even today, with satellites and computerized navigation systems, a Beaufort 0 ("Calm; smoke rises vertically") is the same around the world, while a 12 ("Air filled with foam; sea completely white with driving spray;

visibility greatly reduced") spells disaster no matter how fast the hurricane winds are actually blowing [19].

Musical notation presents another relevant example. Easily transportable, relatively impervious to language, and yet a representation (rather than a reproduction) of a performance, scores have the kinds of conventions that information visualization for the most part still lacks. At this point, good visualizations are featured in "galleries," as befit works of art. They are created by artists and artisans, not by people who merely have something to say. At the risk of a strained analogy, we are at the stage where latter-day monks painstakingly hand-letter sacred texts, still awaiting both Gutenberg and the typewriter.

In his book, *Envisioning Information* [20], Edward Tufte suggested five tactics for increasing information density and "escaping flatland"—conveying more than two dimensions of meaning on paper. These are:

1. Micro/macro readings (relating both wholes and parts as distinct entities);
2. Layering and separation (often by use of color and graphic weight, as in a technical drawing);
3. Small multiples (to show often subtle differences within elements of a system: a good lunar chart is an example);
4. Color and information (sensitivity to the palette as color labels, measures, represents reality, and enlivens);
5. Narratives of space and time (compressing the most powerful human dimensions onto flatland).

For all of the wisdom in these suggestions, and the beauty of Tufte's examples—it is no accident that he is both a statistician and a working artist—good information visualizations remain rare. For information to convey meaning in standard, predictable ways, we need tools: "tools" as in grammars [21] and lexicons rather than more widgets. Somewhat paradoxically, the path to better visualizations will be paved not with software but with words.

References

[1] Tufte, E. R., *The Visual Display of Quantitative Information,* Cheshire, CT: Graphics Press, 1983.

[2] Bertin, J. (translated by W. J. Berg), *Semilogy of Graphics,* Madison, WI: University of Wisconsin Press, 1983.

[3] Lucas, B. D., et al., "Nanoimprint Lithography Based Approach for the Fabrication of Large-Area, Uniformly-Oriented Plasmonic Arrays," *Advanced Materials*, Vol. 20, Issue 6, March 18, 2008, pp. 1129–1134.

[4] Williams, J. G., K. M. Sochats, and E. Morse, "Visualization," *Annual Review of Information Science and Technology (ARIST)*, Vol. 30, 1995, pp. 161–207.

[5] Card, S. K., J. Mackinley, and B. Schneiderman, (eds.), *Readings in Information Visualization: Using Vision to Think,* San Francisco, CA: Morgan Kaufmann, 1998.

[6] Teleu, A., *Information Visualization: Principles and Practice,* Wellesley, MA: A. K. Peters, 2008.

[7] Deitrick, S., and R. Edsall, "Making Uncertainty Usable: Approaches for Visualizing Uncertainty Information," in M. Dodge, M. McDerby, and M. Turner, (eds.), *Geographic Visualization: Concepts, Tools and Applications,* New York: Wiley, 2008, pp. 277–291.

[8] Tory, M., and T. Möller, "Rethinking Visualization: A High-Level Taxonomy," in *IEEE Symposium on Information Visualization*, 2004, pp. 151–158.

[9] Harris, R., *Information Graphics: A Comprehensive Illustrated Reference,* Atlanta, GA: Management Graphics, 1999.

[10] Lengler R., and M. Eppler, "Towards a Periodic Table of Visualization Methods for Management," *IASTED Proceedings of the Conference on Graphics and Visualization in Engineering*, Clearwater, FL, 2007.

[11] Carpenter, S., "Evaluating Information Visualizations," in Kerren, A., et al., (eds.), *Information Visualization: Human Centered Issues and Perspectives,* Berlin: Springer, 2008, pp. 19–45.

[12] Plaisant, C., "The Challenge of Information Visualization Evaluation," *Proceedings of the Working Conference on Advanced Visual Interfaces*, 2004, pp. 109–116.

[13] Kahneman, D., P. Slovic, and A. Tversky, *Judgment Under Uncertainty: Heuristics and Biases,* New York: Cambridge University Press, 1982.

[14] Kosara, R., et al., "Visualization Criticism," *IEEE Computer Graphics and Applications*, Vol. 28, No. 3, May/June 2008, pp. 13–15.

[15] Gastner, M., and M. Newman, "Diffusion-Based Method for Producing Density-Equalizing Maps," *Proceedings of the National Academy of Sciences,* Vol. 101, May 18, 2004, pp. 7499–7504.

[16] Garland, K., *Mr. Beck's Underground Map*, Capital Transport Publishing, 1994.

[17] http://homepage.ntlworld.com/clivebillson/tube/tube.html.

[18] Levinson, M., *The Box: How the Shipping Container Made the World Smaller and the World Economy Bigger*, Princeton, NJ: Princeton University Press, 2006.

[19] Huler, S., *Defining the Wind: The Beaufort Scale and How a 19th-Century Admiral Turned Science into Poetry,* New York: Three Rivers Press, 2004.

[20] Tufte, E., *Envisioning Information,* Cheshire, CT: Graphics Press, 1990.

[21] Wilkinson, L., *The Grammar of Graphics,* 2nd ed., New York: Springer Science, 2005.

Chapter 7

Beyond Visualization: Sonification

Dr. Mark Ballora, associate professor, The Pennsylvania State University

This chapter explores the nature of sound as an informational medium, and considers how it can be effectively employed as a means of monitoring and studying data. Various sound parameters are discussed, as well as their suitability for conveying information in a way that permits meaningful discrimination. The goal is to familiarize readers with ways in which a set of data processing operations may be matched with a set of sonic parameters in order to gain insights into the data's dynamics.

7.1 INTRODUCTION

While the primary task of the sciences may be exploration and the discovery of new knowledge, a critical issue currently facing scientists and researchers is in the area of presentation—the ability to introduce their discoveries effectively, both to laypeople and to fellow researchers. There is an emerging area of interest in representation of scientific information, and how the use of multimedia technologies, an essential component in disseminating new information, can in turn shape and influence scientific thought [1].

The problem concerns dealing not only with new forms of information, but also with unprecedented quantities of it. In our Information Age, new forms of gathering information are constantly being created. However, this does not necessarily lead to increased understanding. In particular, managing crisis situations or monitoring infrastructures requires the ability to interpret incoming information from multiple sources. With new sources of information constantly becoming available, the challenge becomes how to process it effectively, avoiding the condition described by informatics researchers as *cogmenutia fragmentosa* [2].

When the subject of displaying information comes up, or even the term "multimedia," most people are likely to think of some form of visualization. Yet, to risk stating the obvious, as we navigate our way through life, the eyes and the ears play complementary roles in giving us information about our environment. But in research fields, the eyes predominate, as data sets and monitored information are typically presented through visualization. In addition, visual displays are the method of choice to provide a characterization of an emerging situation for applications ranging from military situation awareness, to emergency management to law enforcement. The use of sound, touch, and other senses to convey information has been relatively neglected in all of these areas.

Of course, the ears are not completely neglected as a tool for understanding. Scientists have long used the sonar and the Geiger counter in situational monitoring. Hospitals also use a variety of audio monitors to track patients' vitals, as the repetitious and relatively unobtrusive sound cues quickly retreat to the attentional background when they are unchanging. This allows hospital workers to remain aware of important patient conditions while their eyes are focused elsewhere. There is even historical evidence [3] that Galileo Galilei relied on auditory information to demonstrate the quadratic law of falling bodies: As a ball rolled down an inclined plane, it hit bells that were suspended at uneven distances above the ramp, at increasing distances from each other. As the ball picked up speed on its descent, the bells sounded at regular intervals. The bells were an essential component of this demonstration since timekeeping devices of the seventeenth century were less precise than human rhythmic perception.

The use of nonspeech sound for purposes of conveying information is termed *auditory display*. Auditory display represents a comparatively recent development in the intersection of multimedia technologies and scientific research. It is a broad and general term that encompasses a number of applications. One is the addition of sound elements to graphical user interfaces such as the Macintosh or Windows operating systems to enhance their functionality or ease of use. Another is the addition of sound elements to make such user interfaces accessible to visually impaired users.

Three of the examples given above—the sonar, the Geiger counter, and medical monitors—represent a form of *auditory monitoring*, wherein conditions of interest are continually updated through sound. The last example, Galileo's ramp, represents an early implementation of *sonification*, which explores the use of sound as a means of representing the events under study. While Galileo employed sonification to illustrate the results of motion experiments, in more contemporary implementations, sonification is used to examine data sets. It addresses the question of whether valuable information (which may not be evident with a conventional graphic representation) might become apparent through a sonic representation.

To date, graphical displays serve as the primary medium for presenting data. Various visualization techniques form a vocabulary of commonly used images that are quickly understood [4]. An example is the pie chart, that well-known illustration of proportional subdivisions. Pie charts are common vocabulary, appearing in specialized literature as well as in junior high school-level math textbooks. The 1980s brought tremendous increases in computing power, among them advanced visualization capabilities. Researchers building upon established graphing methods have employed these technologies to create powerful imaging software packages.

In the 1990s, new and inexpensive computer technologies were developed that could generate and process digital audio content. Consumer-level personal computers are now capable of advanced sound signal processing that takes place in *real time* (instantaneously). This has led to a revolution in the music production industry, as evidenced by the myriad of musical "gear magazines" currently in wide circulation. But it has also led a growing number of researchers to explore the use of sound to illustrate and distinguish relative elements of large data sets. Effective use of sound hinges on perceptual understanding and the types of tasks for which we use the eyes and ears.

Visualizations are strongly *synoptic*, that is, an entire image can be seen at once. The eyes provide summary information of features such as shape, size, and texture. In contrast, sonification, like a piece of music, exists in time. It cannot be listened to all at once. Being time-based, the ears give us a strong sense of *dynamic* elements of our environment. The auditory system is also highly adapted for following multiple streams of information [5]. That is, listeners can readily apprehend a number of simultaneous melodies if they are presented effectively. Thus, sonification is an effective way to display a multitude of signal processing operations simultaneously, with each being represented as a line of counterpoint, a series of chords, or a succession of musical instruments. However, while visualization has a recognized vocabulary of imagery, sonification does not—there is no auditory equivalent of the pie chart.

The development of auditory displays is an inherently multidisciplinary activity. A successful auditory display must combine elements from perceptual psychology, music, acoustics, and engineering [4]. Auditory displays, then, are best realized in an interdisciplinary environment, with sound specialists who possess a working knowledge of the research area working alongside researchers who have a working knowledge of sound realization.

This chapter explores the nature of sound as an informational medium, and considers how it can be effectively employed as a means of monitoring and of studying data. Various acoustic components of sound are discussed in terms of their suitability for conveying information in a way that permits meaningful discrimination. The goal is to familiarize readers with ways in which a set of data processing operations may be matched with a set of sonic parameters in order to gain insights into the data's dynamics.

7.2 SOUND AS INFORMATION

Current efforts towards advancing the use of sound to convey information have been largely due to the efforts of the International Community on Auditory Display (http://www.icad.org). The precedents set by their work are foundational to an understanding of auditory display. Their seminal publication *Auditory Display: Sonification, Audification and Auditory Interfaces* [5] is a collection of papers taken from the first conference in 1992, which defined the field and its objectives of exploring the uses and potential for conveying information through sound in technology.

The distinction between the terms *sonification* and *audification* is defined in Gregory Kramer's introductory survey [4]. He suggests that the term *audification* be used in reference to "direct playback of data samples," while the definition of *sonification* is taken from Carla Scaletti's paper to refer to "a mapping of numerically represented relations." This distinction, presented in 1994, is still in use in the ICAD literature and will be employed in this overview.

The term *mapping* will appear throughout this chapter to refer to the translation of information to illustrative elements. While mapping of information to visual elements has an established canon of techniques in the field of visualization, auditory mapping is still in its formative stages. The basic concept of transforming data into sound representations is shown in Figure 7.1. The steps range from acquiring basic information in a time-series format to parsing and filtering (including special transformations such as time-domain and frequency-domain transformations), followed by further parsing and sound design. At this point this process is a creative one in which techniques are selected based in part upon the nature of the underlying signal data to be represented and in part upon creative design to represent patterns of data, such as anomalies, in special ways that alert a user/analyst.

7.3 MONITORING VERSUS ANALYSIS

As described in Section 7.1, a number of real-time auditory monitoring implementations are in common use. In general, the goal of monitoring is to highlight known conditions. Normal conditions must be easily distinguishable from a set of known conditions that trigger some sort of alarm signal. Monitoring also tracks conditions as they are occurring, in real time.

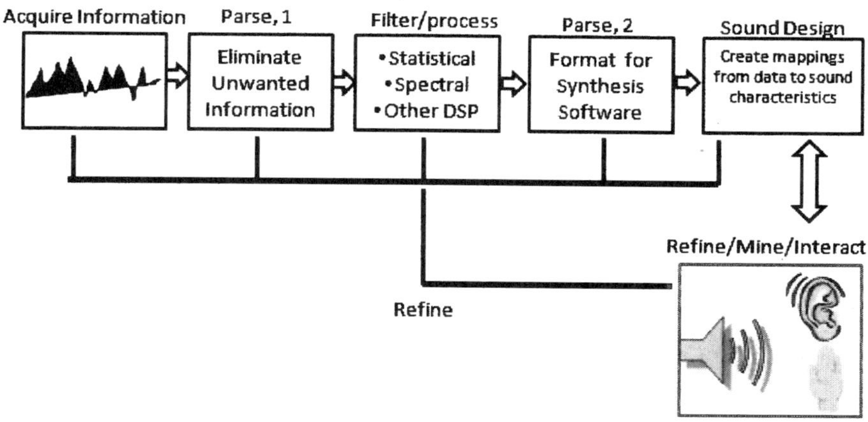

Figure 7.1 Concept of transforming data into sound representations.

Beyond monitoring, using sound as a component of data analysis brings up new problems. An analytical illustration system must have enough flexibility to allow unknown conditions to emerge. An analytical system does not exist in real time, but rather is something that is studied after the fact. This nonreal-time nature allows great flexibility in time resolution. Great volumes of data can be compressed to whatever playback time is desired.

7.4 ACOUSTIC DIMENSIONS OF SOUND

This section examines the dimensions of sound events and is meant to provide an introductory acoustic vocabulary.

7.4.1 Air Pressure

Sound events occur due to changes in air pressure (the density of air molecules). When air molecules are disturbed by some event, such as hands clapping, they are squeezed together, bounce back apart, and then bounce back together again in a repeating series of compressions and rarefactions from undisturbed pressure levels. Sound wavefronts travel at a rate of 344 m/sec (1,128 ft/sec), with small variations introduced by temperature, humidity, and elevation. Incidentally, this is the same property commonly discussed by weather forecasters, with the difference that atmospheric pressure changes are at a much greater magnitude and much lower frequency than acoustic pressure changes [6].

The energy of a sound event expands outward from the location of the event, eventually encountering our eardrums, which move inwards and outwards in response to air pressure changes. The motion of the eardrums is transduced into electrical impulses in the inner ear, which the brain interprets as sound.

7.4.2 Pitch/Frequency

Periodic (repeating) variations in air pressure result in sounds that are pitched, while random air pressure vibrations result in noise, which has a much less pitched quality (if any). Noise can be made pitch-like if it is put through narrow bandpass filters, which produce a quasi-pitched whistling sound.

Changes in pitch correspond to changes in frequency, with higher frequencies producing higher pitches. Pitch is given names corresponding to letters (A, B, C . . . G) or solfège syllables (*do, re, mi . . .ti*) by musicians. Frequency is measured in cycles per second or hertz (Hz). *Wavelength* (λ) is the inverse of frequency and is obtained by dividing the speed of sound by the frequency. The human ear, on average, can perceive frequencies within the range of 20 Hz ($\lambda = 50$ ft) to 20 kHz ($\lambda = 0.05$ ft). As a point of reference, the pitch middle A usually corresponds to a frequency of 440 Hz (although musicians sometimes tune to a slightly different frequency for A, such as 442 or 444).

7.4.3 Volume/Amplitude

A vibrating system's amplitude is the degree of displacement from equilibrium. Acoustic pressure changes are measured in pascals (or newtons per square meter). Greater average pressure levels sound louder. An oscilloscope-like view (also termed time-domain view) of a sound wave illustrates changes in air pressure as a function of time.

There are two difficulties in measuring acoustic amplitude levels. One is that the range of human hearing spans many millions of pascals. The other is that sound pressure levels are relative, not absolute. Air molecules are never completely motionless, so there is no state of zero pressure.

To address these difficulties, acoustic amplitude levels are measured on a scale that is comparative and logarithmic. The decibel (dB) scale, compares a given sound's pressure level, p, with a threshold level (the lowest audible pressure level of a sine tone at 1 kHz). The threshold has been set at 2 × 10 minus 5 pascals (p_0). The pressure level of a given sound may be represented in decibels by the equation:

$$L_{(dB)} = 20\log_{10}(p/p_0)$$

The decibel scale is comparative in that it compensates for the absence of a zero level by comparing (through division) any given level with the threshold. Taking the logarithm of that division reduces the wide range of audible pressure changes into a more manageable range. Halving a pressure level lowers it by 6 dB; doubling a pressure level raises it by 6 dB.

7.4.4 Timbre/Spectrum

Timbre is the perceived difference in sound quality that distinguishes two different instruments (e.g., a violin and a flute) playing at the same pitch and loudness.

Sinusoidal sound waves are called *pure tones*, because they vibrate at only a single frequency. Natural objects, in contrast, vibrate at a multitude of simultaneous frequencies. Timbre corresponds largely to spectral content. Periodic sounds, just like any periodic wave, can be processed by a Fourier transform, which yields a *spectral* view of the wave (in contrast to the time-domain view provided by oscilloscopes). A sound's spectrum gives information about its component frequencies, their relative amplitudes, and their relative phases. The simultaneous frequencies in a vibrating object are called its *partials*. As demonstrated by the Fourier theorem, the partials of periodic (repeating) vibrations consist of *harmonically* related partials, that is, partials that are integer multiples of the lowest partial, the *fundamental*, which determines the perceived pitch.

Timbre, however, defies simple explanation. Though its spectrum is largely responsible for a sound's timbre, sounds tend not to be spectrally consistent, but to change over time, both in loudness and in spectral content. The amplitude shape of a sound over time is called its *envelope*. The envelopes of musical instruments, as an important example, begin with an initial *transient,* or *attack,* portion. Following the transient, the volume level stabilizes as instruments usually fall into a *steady-state*, or *sustained*, vibration.

The transient is characterized by many high frequencies and noise. Examples include the scraping of a bow or the *chiff* of breath on a flute. An instrument's distinctiveness is determined primarily by the transient portion of its sound. If a recorded note has its transient removed, it can be very difficult to identify the instrument. While changing the relative phases of a steady-state sound has no audible effect (although the waveshape may change significantly), altering the relative phases of frequencies appearing in the transient can produce audible timbral effects. In contrast, the steady state is characterized by both periodicity and a harmonic spectrum and creates the pitched portion of a note.

7.4.5 Location/Stereo Pan

The ability to localize auditory objects is based on numerous cues, some physical, some learned. There are three primary physical cues: *interaural time difference* (ITD), *interaural intensity difference* (IID), and spectral difference.

Wavelengths that are larger than the dimensions of an obstacle diffract around it. Thus, a sound wavefront occurring to one side will reach the nearer ear and then diffract around the head and encounter the farther ear. Interaural time difference is the time delay between the wavefront reaching the two ears. This is the most powerful localization cue for frequencies under 1,500 Hz. It is also called the *precedence effect* or the *Haas effect* in sound reproduction contexts. With an identical sound stimulus emanating from multiple loudspeakers, all of which are at different distances from the listener, listeners will tend to localize the sound at the nearest loudspeaker, which produces the wavefront that reaches the ear first. A more distant speaker needs to be over 18 dB louder than a nearer speaker to overcome the localization effects of the first wavefront.

When wavelengths are smaller than the dimensions of an obstacle, they are reflected back the way they came. Thus, higher frequencies do not diffract around the head, but are reflected away from it. Frequencies above 1,500 Hz have a wavelength under 21 cm, the average diameter of the human head. These higher frequencies tend to reflect off of the head, resulting in IID, an *acoustic shadow* at the farther ear.

The perception of elevation is due to reflections of the wavefront off of the shoulders, as well as filtering carried out by phase differences due to reflections off the pinnae. This filtering provides the spectral cues that give information about elevation.

In describing the perceptual system's treatment of location, [7] quantifies its tendencies with the term *localization blur*, which is a measure in degrees of the average margin of error present in a given region. In the sanitized conditions of a laboratory, where stimuli are tightly controlled and limited to pure tones, clicks, noise, and speech samples, the minimum localization blur in any direction has an average near 1°. In this regard, the auditory perceptual system demonstrates less resolution than does the visual system, with which changes in position have been perceived at less than 1 minute of an arc.

Perception of direction is most sensitive in a forward, horizontal direction (also known as the lateral field), with 0° being the direction in front of the listener's nose. The localization blur increases as sound sources move away from this area. At ±90°, the localization blur is three to ten times as great as at 0°. Sideward localization accuracy drops according to a *cone of confusion*, which refers frontward sounds being difficult to differentiate from rearward objects by the same degree factor. Imaging reconsolidates towards the rear. The localization blur of objects directly behind averages twice that of the forward perception.

Elevation perception is less certain. Elevation tests involving continuous speech of an unfamiliar voice have shown a localization blur of 17°, a blur of 9° when the speech is that of a familiar voice, and 4° for wideband noise. With the stimulus of narrowband noise, there is virtually no perceptibility in elevation; instead, the perception of height becomes associated with the pitch of the sound. The higher the pitch, the higher in height is the sound's location perceived.

Audio engineers simulate localization via loudspeakers by creating intensity differences. While ITD is by far the stronger cue, its effectiveness is dependent on the listener being in a central "sweet spot," equidistant from each speaker. The effectiveness of intensity panning, on the other hand, can be appreciated within a much wider listening area. It is only the rare audiophile who sits stationary in a central listening position when listening to music at home. For this reason, intensity panning, rather than delay panning, is employed in the vast majority of commercial recordings. To a blindfolded listener, a signal from two equidistant loudspeakers will be localized in space, directly between them, a *phantom image*. Changing the intensity of one speaker "pulls" the phantom image toward the louder source. A gradual volume cross-fade between two speakers will make the sound appear to move from one to the other location.

More specific localization images can be obtained by simulating the filtering done by the pinnae. Attempting to create such effects is problematic for two reasons. One is that each individual's pinnae produce a different filtering operation. Researchers have had some success through the use of *head-related transfer functions* (HRTFs), which are a general model of a typical ear's response. Effects through HRTFs are very dependent on listener location, however, and are most effective when played over headphones, or in close listening environments such as personal computer speakers.

7.4.6 Distance

As an object approaches or recedes, there is an amplitude change, as well as a change in the ratio of *direct* to *reverberant* sound. As the distance between a sound event and the listener increases, the wavefronts emitted directly by the object reach the listener at quickly decreasing levels, according to the inverse square law. Even at short distances, the level of direct wavefronts may be quite low in comparison with wavefronts resulting from energy reflected off of walls and furniture. As objects move towards the back of an auditorium, virtually all sound is likely to be due to reflections and the resultant reverberation, with little to no direct sound. Different frequencies are "favored" in different rooms, with the reverberation functioning as the "signature" of the space. When sound events occur outdoors, where there are no walls or ceiling to reflect them, as an object moves away from a listener, it quickly drops in volume as well as in high-frequency content, as high frequencies are more quickly absorbed into the air than lower frequencies.

7.4.7 Duration

Long unchanging events, lasting tens of seconds, function as drones. For events that occur between 16 Hz and 0.125 Hz (that is, over time spans ranging from 62.5 milliseconds to 8 seconds), a rhythm is perceivable. For extremely short events that are on the order of milliseconds, the effect is a sound cloud, in which individual events are not perceived, but an overall impression is created by many hundreds or thousands of short events each second.

7.4.8 Meter and Harmony

While these two terms are based on musical techniques, they are readily perceptible to nonmusically trained listeners. Most listeners, for example, can distinguish between a tango and a waltz, even if they lack the musical vocabulary to describe them. An understanding of what characterizes musical styles adds a myriad of possible qualitative shifts.

7.5 PSYCHOACOUSTICS: DIMENSIONS OF HUMAN SOUND PERCEPTION

A sonification maps numeric information to sound events, with characteristics of the data determining characteristics (or dimensions) of sound events. However, the human auditory system does not respond uniformly to changes in each of the acoustic dimensions described in Section 7.4. What reaches the brain as perceived sound events are not objective representations of acoustic information, but rather interpretations of them.

7.5.1 Pitch Perception

The human auditory system perceives pitches on a logarithmic scale, such that every doubling of frequency represents a duplication of pitch class—that is, a jump from *do* to a higher (or lower) *do* or from *mi* to a higher or lower *mi*—a phenomenon that musicians refer to as a change in octave. The consistent identity of the various pitch classes can be demonstrated by sitting at a piano and playing a familiar tune such as "Happy Birthday" and randomly transposing each note of the melody by some number of octaves up or down. Though the pitches sound higher or lower than those used by singers, the overall contour of the melody remains intact, and the song is immediately recognizable (although pitches tend to lose their identities at frequencies below 40 Hz and above 5 kHz).

The logarithmic nature of pitch perception can be further understood if we consider the nature of the Western tuning system of *Twelve Tone Equal*

Temperament, which is based on equal perceptual subdivisions of the octave. An octave may be described mathematically by choosing a starting frequency and multiplying it by

$$2^{n/12} \qquad \text{for } n = 0, 1, 2, \ldots 11$$

For example, a starting frequency of 220 Hz would yield the following set of pitches: 220, 233, 247, and so forth. These are illustrated in Figure 7.2, showing the subdivisions of an octave on a piano.

When two tones are close in frequency, it may be difficult to hear the difference between them when they are played in succession. However, when they are sounded simultaneously, there is an audible beating at the difference frequency between the two tones, due to cancellations and reinforcements that occur when the two waves interact. For example, if tones at 440 Hz and 444 Hz are played, there is an audible beating at 4 Hz. If the difference in frequency is increased, the beating frequency increases accordingly, to the point that the beating sensation is perceived as a roughness. At a certain point, the roughness disappears and the perception is that the roughness smoothes and what is heard is two discrete tones. This point of transition is called the *critical band.*

When harmonically related frequencies are combined, the discrete frequencies "fuse" so that they are not heard as separate pitches, but rather as a pitch with a complex timbre. With training, people can learn analytic listening, wherein they can "hear out" individual harmonics of a complex tone. In

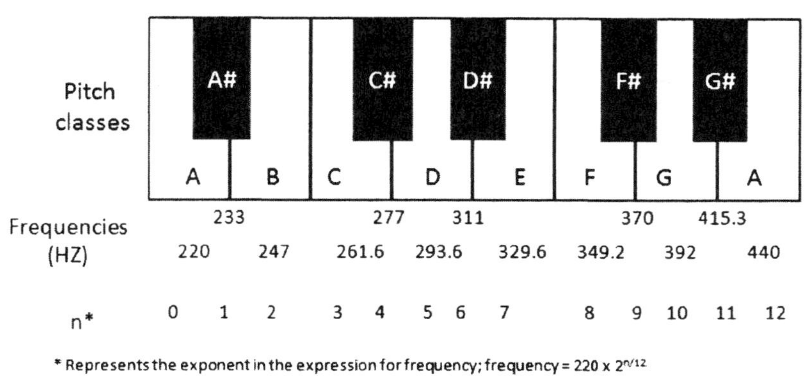

Figure 7.2 Piano scale subdivisions of the octave.

single contrast, when inharmonic tones are played simultaneously, they retain their individuality and do not fuse to the same degree. Under some circumstances, the auditory system will interpret a fundamental that is not actually present. For example, if the tones 100 Hz, 150 Hz, 200 Hz, and 250 Hz are combined, many listeners will report hearing a *missing fundamental* tone at 50 Hz.

7.5.2 Loudness Perception

Like pitch, amplitude is also measured on a logarithmic scale. However, this is a convenience and is not based on the same level of perceptual duplication that is associated with pitch. While doubling a frequency yields a concrete duplication of pitch class, doubling pressure levels does not necessarily mean that the volume will be twice as loud.

While loudness is primarily related to amplitude levels, there are a number of factors that contribute to the perceived loudness of a sound event. One factor is frequency. Equal Loudness Curves (created by Fletcher and Munson in the 1930s) show that pure tones played at the same decibel level over a range of frequencies are not perceived as being at the same loudness level. Very low and high frequencies sound softer than frequencies in the range of 1–5 kHz (frequencies that tend to be components of human speech). It is for this reason that some hi-fi amplifiers include a loudness knob, which is meant to give an extra boost to bass frequencies, which will tend to be less audible at low volume levels.

Loudness also drops when two sources of the same frequency are combined. Two violins playing different pitches generate twice the acoustic energy of a single violin, and they generate twice the volume as a soloist. But if they play the same pitch, they do not sound twice as loud, although they generate twice the acoustic energy of a single violin. As a rule, to double the volume level of instruments playing the same part, eight violins are required. Put another way, perceived loudness is proportional to the cube root of acoustic energy when multiple sources sound at the same frequency.

Perceived loudness is also dependent on its *bandwidth* (component frequency content). Increasing the bandwidth of a sound makes it appear to sound louder, even if its acoustic energy remains constant.

Other scales besides the decibel scale have been proposed in an effort to measure perceived volume levels, such as the *phon* and *sone* scales, but these are only informative under certain circumstances. Despite many efforts, no one has succeeded in creating a definitive perceptual scaling system for loudness. Loudness may be considered an *emergent* property, that is, one that is the result of many factors, and that is not always predictable.

7.5.3 Timbre Perception

Timbre is also an emergent property. While it is primarily related to a sound's spectrum, it is also dependent on a number of other factors. As a starting point, research in sound synthesis has shown that to create effective imitations of instruments, a match of envelope shape is more important that an exact match of its spectrum. The attack portion is also critical. As stated in the previous section, the nature of an instrument's transient is critical to its signature. Another component of the transient is that a faster attack can be confused with brightness (more high-frequency overtones). With the advent of computer music synthesis, many studies [8–10] have proposed quantitative classifications of timbre based on overtone content, envelope shape, and attack time. However, like loudness, there are no perfect perceptual measurements of timbre.

7.6 AUDITORY GESTALTS

Sonification involves synthesizing sound events in order to create an informative aural impression. Thus, an effective sonification design should be informed by knowledge of how the auditory system creates a description of the *auditory scene*, or how the single complex wave that reaches the eardrum is parsed by the auditory system to a number of apparent sound sources. How is it, for example, that a visually impaired person can enter a busy aural environment such as Grand Central Station, and be able to find information, exits, or the desired train platform? Gestalt psychologists, beginning with Max Wertheimer in 1912, have identified rules of visual organization. These rules describe how our senses innately give structure to information whenever possible, perceiving not individual elements, but an organized whole. The classic example is that of a movie marquee with light bulbs around its edges that shine in succession, creating the effect of a moving light. Though we know that no light is actually moving, we cannot help but see that effect due to our tendency to organize a gestalt out of the series of individual lightbulb activations. Gestalt literature identifies grouping principles based on similarity, proximity, good continuation, habit/familiarity, belongingness, common fate, and closure.

Auditory psychologists such as A. S. Bregman have applied these same principles to audition and identified methods by which the auditory system identifies *auditory streams*, or apparent sources of sound events. The examples on the CDs that accompany [11, 12] provide many compelling examples. A musical passage loses all coherence when alternating notes are panned to opposite channels; a quick sequence of tones sounds like a single line when all frequencies are similar, but if alternating tones sound in two consistently separate frequency ranges, two apparent lines are heard; or a single complex tone becomes two tones if vibrato is applied to a subset of its component frequencies.

7.7 SONIFICATION: DESIGN CONSIDERATIONS

Creating effective sonifications involves synthesizing discrete auditory streams in a way that is both engaging and informative. The most common strategy is *parameter-based mapping*, wherein each data point becomes the basis of a sound event, the characteristics of which are dependent on the data value. Data values are fed into a music synthesis program to generate a sequence of sound events, with acoustic dimensions determined by the data value. For a one-dimensional data set, each data value can be made to apply to as many parameters as desired (frequency, volume, stereo pan position). Multidimensional data sets may have different dimensions mapped to different parameters.

Fundamental to the auditory illustration is the *timescale,* the duration of events and the time between them, which, as described earlier, creates a texture along the axis of drones to sound clouds. Sound clouds are a musical technique that has been explored by composers such as Gyorgy Ligeti (1923–2006) and Iannix Xenakis (1922–2001), both of whom wrote for large ensembles of instruments with many separate lines. Xenakis wrote extensively on techniques of *stochastic music*, whereby large numbers of sound events are generated according to laws of probability, and the overall effect is akin to raindrops or a swarm of insects [13]. In the area of computer music, granular synthesis [14] is an area that concerns itself with generating large densities of short sound objects, analogous to scatter plots or to the pointillist paintings of artists like Georges Seurat (1859–1891), in which images were composed of hundreds of small dots.

At a minimum, a successful sonification development environment must allow parameters to be kept perceptually separate. In a "multitrack" sonification, streams of information from different data sources can be presented with an interface akin to an audio mixer, whereby different instruments may be mixed at higher or lower levels, depending on what type of impression is most helpful to the user. This approach has the flexibility of allowing other elements to be added to the mix, allowing an arbitrary number of processing operations to be heard in tandem.

As is the case with zooming in or out on an image, the ability to change the playback rate brings out varying levels of abstraction, depending on the degree of time compression employed. Users must also be able to begin rendering sound from any arbitrary point in the data set. Finally, there must be a clear distinction between primary cues, which are meant to illustrate moment-to-moment changes in the data, and secondary (or supporting) cues—these are sound dimensions to which we are less sensitive, and which may be perceptually intermingled in some cases. They are less useful for tracking moment-to-moment changes, but can be quite helpful for distinguishing among different auditory streams (or data dimensions).

An interesting design consideration involves the polarity [15] of a realization. Polarity represents how changes in a data property are mapped along a particular sonic dimension. For example, should a condition of increased urgency be mapped to an increase in tempo (a positive polarity) or a decrease in tempo (a negative polarity)? Different studies are likely to require different types of representation.

In visualizations, height often means "more," a greater magnitude of some kind. Pitch is a natural sonic correlate, such that a higher pitch can signify greater magnitude. The use of pitch involves relative changes. Only the rare individual who possesses perfect pitch would be able to identify the numerical value of a sounding frequency. However, fluctuations in pitch are adequate to indicate relative changes in value. The human ear is highly sensitive to changes in frequency, such that even small changes are readily perceivable as differences in pitch. Thus, changes in a data set may be magnified if they are mapped to a pitch. A change that is not easily perceptible when mapped to a change in a line's direction may be more readily perceived when mapped to a change in pitch. Pitch, then, may be considered a primary cue, as it is effective for representing small changes on a moment-to-moment basis. Walker [16] raised the issue of polarity in pitch mappings, observing that while higher pitches might seem suitable for representing greater temperatures, they are also suitable for representing smaller, with lower pitches representing larger. (This polarity has the real-world correlate of objects such as bells.) The suitability of a polarity and pitch range will depend on the type of information being conveyed and is also subject to listener preference.

Volume is another possible magnitude correlate, although this parameter is problematic due to the difficulty of assigning definite loudness scales, as described above. Given the ambiguity of loudness as a precept, this is a mapping that is likely to be most effective in measuring changes on a large scale, perhaps in tandem with other parameters. It is less likely to be effective in conveying small-scale changes in magnitude.

Frequency, then, would be the preferred method to convey magnitude, although due consideration must be given to the size of the changes involved. Due to the logarithmic nature of the auditory system's pitch perception, also described above, changes in the lower-frequency ranges of only a few cycles per second can produce differences on the order of a number of musical scale steps, while much larger changes in frequency are required to produce the same relative pitch change in higher ranges. Hence, it is typically preferable to map changes of frequency on a logarithmic, rather than a linear, scale. This will ensure that changes in the data are reflected in equal changes of musical pitch interval. Depending on the equation put to use, the mapping can produce any pitch range. If all data points fell into a range, for example, of 0.0 to 1.0, then each data point could be mapped into an exponential equation something like the one used above to describe frequency values in equal temperament. The resulting frequency range would then be between 2^0 and 2^1, or the span of a musical octave. Depending on the range of the data set in question, the mapping equation can be easily adjusted to produce frequencies within a desirable pitch range.

Another possible aspect of volume is the characteristic of *tremolo*, or *amplitude modulation*. As discussed above, this results when two sounds that are near to each other in frequency are juxtaposed. This can be very effective in representing characteristics that are close to an equilibrium, or proximity. As the balance drifts, or the distance lessens, the beating (or absence of it) can represent the degree of proximity or distance. In another implementation, a synthesized tremolo rate may be mapped to some data parameter. Changes to both the pitch of a sound and the tremolo rate can be highly recognizable (indeed, both rely on the ear's ability to detect periodicity). Thus, a sound may be presented that has both varying pitch and tremolo rates, which may be termed *orthogonal* precepts, as the two effects can be perceived simultaneously without being confused with each other.

Color, or brightness, is often an important component in a visual display to differentiate between different types of elements. A literal mapping of color to the auditory domain might involve pitch, since both color and pitch are related to frequency. But if pitch is best employed to represent changes in magnitude, then perhaps a more suitable correlate for color is timbre. Indeed, musicians often informally refer to timbral characteristics as color, with comments such as "This piece brings interesting colors out of the piano." It is unlikely, however, that small changes in timbre could be an effective basis for an auditory measurement. Like loudness, timbre is probably best employed to reflect large-scale changes, or as an enhancement in combination with other parameters.

Timbre and volume, then, are supporting parameters. They may be used effectively in conjunction with another parameter to give it greater distinction. For instance, as a sound's tremolo rate increases, it might also be given higher harmonic content and greater volume, so that a faster tremolo sounds "buzzier" and louder.

Timbre may also be used to differentiate among the different data tracks, so that they function as separate instruments. Different data dimensions may be assigned to timbres that are meant to blend harmoniously, but that remain distinct if the listener focuses on them. Thus, a multitrack sonification employs the "cocktail party effect," a term that refers to the attentional filtering the auditory system is able to carry out, enabling us to focus on one speaker's voice in a crowded room [17].

Another possible component is that of location. Like volume and timbre, this is best used as a supporting parameter, as it is less effective for mapping small changes in data. Localization, however, is not a simple cue. Bregman [11] observed that localization alone is not sufficient as a means to discriminate independent auditory streams. In life it is rare that we hear only a direct sound source; enclosed spaces, surfaces, and obstacles all create a multitude of reflections. Thus, all identification of objects through hearing would break down if each reflection were indicative of a new auditory event. However, we get a great deal of information from the timbral changes introduced by these multiple reflections. The superimposition of the sound wave with copies of itself creates reinforcements or cancellations of certain frequency regions, an effect known as *comb filtering*. As is the case with small frequency differences, the auditory system is highly sensitive to small differences of inter-onset time. This sensitivity is used to assess acoustic environments and reverberation. It would appear that the evolutionary process has been carefully selective in how our perception of location has developed. For example, differences in phases of a complex tone do not change the tone's primary characteristics. If the tone is steady, introducing phase differences will have at best a minimally audible effect. However, phase differences experienced as inter-onset times of sound events, either among overtone components during the attack portion of a sound event or as reflections of a sound as a component of reverberation, give qualitative information about the listening environment.

Early papers on stereophony in music recording noted that the effect of adding channels was not so much the ability of the listener to perceive precise apparent locations of instruments, but rather a more qualitative impression of *spaciousness* [18]. While listening to music through one speaker, the impression was that of hearing through a window the size of the speaker; listening to music through two speakers gave the impression of an elongated window that filled the space between the two speakers. Bregman suggested that readers experience the effects of stereophony by covering one ear in a concert hall. He noted an increased level of segregation among the various instruments, a factor that audio engineers call *transparency*. Indeed, one characteristic of preferred concert halls is that they exhibit low interaural cross correlation, that is, the hall's reflections should ensure that the wavefronts reaching the two ears should be different, not similar.

Localization can also be highly effective when used in conjunction with other parameters. Two tones close to each other in frequency may be indistinguishable if heard monophonically over headphones. Simulating spatial separation via interaural intensity difference, however, can cause the two tones to segregate and be perceived as two separate pitches [19].

7.8 EXAMPLE SONIFICATIONS

Developmental work in sonifications and auditory displays has proceeded along two broad areas of focus. The first focus area is *foundational* and deals with establishing general principles and best practices [15, 16, 20]. The other focus area is in *applications*, which involves finding answers to specific research questions and problems. The two focuses have a chicken-and-egg relationship. It is unlikely that sonification will be considered a significant research tool without being shown to help solve a particular problem, and de Campo [21] pointed out that it is very difficult to create a generic sonification—different types of problems call for different types of solutions. However, it is also unlikely that a critical mass of people will be interested in exploring sonification as a research tool if there are no established principles and guidelines to build on.

Successful applications of sonification are few, but they are promising and clearly worth further investigation. Proof of concept has been achieved in a number of areas, though this has not necessarily led to further development.

Kramer [22] presented a multivariate representation of financial data, assigning sound parameters to properties such as closing figures from different financial exchanges, bond prices, and the value of the U.S. dollar. This paper was written some years before the advent of day trading, in which investors (or would-be investors) buy and sell stocks quickly via special software packages. Day traders make decisions on a minute-by-minute basis, tracking various indices when timing a decisive mouse-click to buy or sell. In 2006, Edward Childs and Stefan Tomic obtained a patent for a sonification system and methodology based on their work developing sonifications for the finance industry (U.S. Patent 7135635) [23].

Researchers at UC Berkeley used sonification to detect quantum interactions [24]. Quantum mechanics equations have long predicted particle current oscillations between two weakly coupled macroscopic quantum systems, although these oscillations had never been observed. These researchers used two reservoirs of a helium superfluid. Membranes in the reservoirs traced voltage changes. Oscilloscopes revealed nothing useful in terms of the oscillations between the two reservoirs, but when the voltage was audified, a clear tone emerged that revealed the expected oscillations. Further observations were then carried out through the study of sound recordings of these tones.

Sonification has provided promising results in medical informatics. Sonifications of heart rate variability [25] and EEG readings of brain activity [26] have been shown to have diagnostic potential. These sonifications have subsequently proved useful as general introductions to physiological health, making distinctions between healthy and diseased states easy for uneducated listeners [27].

Along with using sonification to study data, another potentially valuable implementation is the use of sound as a real-time feedback training device. Wallis [28] described a sonification system designed to function as a component of an immersive stroke rehabilitation system, where physical therapy patients learning to correct sensorimotor deficits have their movements represented as sound so that they can "hear" their movements, and by recognizing certain patterns, they can re-learn simple motions such as reaching for objects and grasping them. In another example, Robert Grober has developed a Sonic Golf Club (http://www.sonicgolf.com/) that gives golfers audio feedback about the nature of their golf swing so that they can learn to improve it. These examples have the benefit of a well-defined outcome (learning to develop a physical activity) and a clear advantage gained from the use of audio information—users' eyes can remain focused on their primary task while they get immediate feedback from another sense.

A number of collaboratives have been formed to explore sonification projects. The Sonification Lab at Georgia Tech's School of Psychology (http://sonify.psych.gatech.edu) is active in developing sonification-based work. SonEnvir (http://sonenvir.aut), hosted at IEM Graz in Austria from 2005 to 2007, was a cooperative of four universities that explored sonifications of sociological data and theoretical physics. Their work was considered promising by many participants, but a shortcoming cited was that nowhere was there a true "aha!" moment that made the sonifications truly indispensable. A conclusion of the work was that the value of the sonification work is in the social processes it requires, rather than in specific outcomes. The precise nature of interdisciplinary collaboration is enough of a framework to bring about interesting results that would not arise from the current paradigms.

7.9 APPLICATION TO CYBERSECURITY

The authors, Ballora and Hall, have conducted initial experiments using sonification to assess the dynamics of cybernetwork traffic [29]. In collaboration with researchers from the University of Buffalo, Hall, Hellar, and McNeese developed a prototype system to process software sensor reports concerning network traffic and reports about activities such as denial of service attacks. This activity was displayed on a cyber situation assessment display along with statistical information about types of reports.

The preliminary results of using sound to characterize network activity are promising. We found that certain types of network attacks provided distinctive and readily recognizable sounds indicating anomalous behavior. Current research is focusing on sonification of network data associated with particular IP addresses (see Figure 7.3).

7.10 SONIFICATION SOFTWARE

A number of sonification software programs have been written. The Sonification Sandbox from Georgia Tech's Sonification Lab (http://sonify.psych.gatech.edu/research/auditorygraphs/sandbox.php) allows data values to be mapped to MIDI

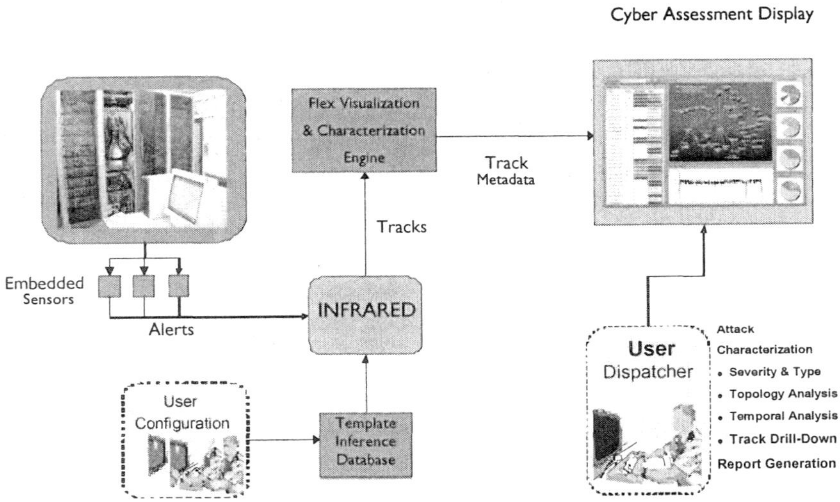

Figure 7.3 Example of cyberattack alert system [29].

values that determine pitch, volume, and timbre. It is easy to use and is a good tool for introductory work in sonifying data sets. However, it is limited in its capability. More complex sonifications are typically done on higher-end synthesis or analysis programs such as MATLAB (http://www.mathworks.com/products/matlab), Mathematica (http://www.wolfram.com/), SuperCollider (http://supercollider.sourceforge.net/), Max/MSP (http://www.cycling74.com), or Pd (http://puredata.info).

7.11 CONCLUSIONS

Given the lack of universal acceptance of sonification systems to date, one could understandably elect to disregard sonification as lacking potential. On the other hand, as researchers and analysts currently face data sets of higher dimensions or multiple simultaneous situation reports from different areas, there is a commonly acknowledged problem of information overload. Therefore, one might reasonably take the position that sound must become a component as information systems take on greater complexity. The question is not whether auditory displays are worth pursuing; it is rather how the utilization of them can be improved. The fact that no "aha!" moment has yet caused their universal acceptance does not mean they should not be explored. For the increasing tsunami of anticipated data related to the human landscape (namely, human-generated data about human activities), sonification represents the potential to address very large data sets and to use background sounds to convey normal or typical behavior. When an anomalous condition appears, the sound changes could direct one's attention to a situation display using more conventional visual situation displays.

Sound is an essential component to any multisensory presentation. To test this, turn down the sound of a TV show or movie, leaving on captions that allow the dialogue to be visible, so the plot can be followed. Then ask yourself, "Is this as effective as it is with the sound up?" Auditory information is critical to creating an effective understanding of our environment. How can it not merit equally important consideration when designing information display systems?

References

[1] *Visualizing Science: Image-Making in the Constitution of Scientific Knowledge (an interdisciplinary symposium),* October 24, 2007, Brandeis University, http://culturalproduction.wikispaces.com/visualizing_science.

[2] McNeese, M. D., and M. A. Vidulich, (eds.), *Cognitive Systems Engineering in Military Aviation Environments: Avoiding Cogmenutia Fragmentosa,* Dayton, OH: Wright Patterson Air Force Base, CSERIAC Press, 2002.

[3] Vogt, K., et al., "Sonification of Spin Models: Listening to Phase Transitions in the Ising and Potts Model," *Proceedings of the 13th International Conference on Auditory Display,* 2007, pp. 258–265.

[4] Kramer, G., "Some Organizing Principles for Representing Data with Sound," in *Auditory Display: Sonification, Audification, and Auditory Interfaces*, G. Kramer, (ed.), Reading, MA: Addison Wesley, 1994.

[5] Fitch, T., and G. Kramer, "Sonifying the Body Electric: Superiority of an Auditory over a Visual Display in a Complex, Multivariate System," in *Auditory Display: Sonification, Audification, and Auditory Interfaces*, G. Kramer, (ed.), Reading, MA: Addison Wesley, 1994.

[6] Sundberg, J., *The Science of Musical Sounds*, 1983. New York: Henry Holt and Company.

[7] Blauert, J., *Spatial Hearing: The Psychophysics of Human Sound Localization,* rev. ed., Cambridge, MA: MIT Press, 1997.

[8] Moorer, J. A., and J. Grey, "Lexicon of Analyzed Tones (Part I; A Violin Tone)," *Computer Music Journal,* Vol. 1, No. 2, April 1977, pp. 39–45. Moorer, James A., and Grey, J., "Lexicon of Analyzed Tones (Part II; Clarinet and Oboe Tones)." *Computer Music Journal* 1(3) (June 1977): pp. 12–29.

[9] Gordon, J. W., and J. M. Grey, "Perception of Spectral Modifications on Orchestral Instrument Tones," *Computer Music Journal*, Vol. 2, No. 1, July 1978, pp. 24–31.

[10] Wessel, D., "Timbre Space as a Musical Control Structure," *Computer Music Journal*, Vol. 3, No. 2, June 1979, pp. 45–52.

[11] Bregman, A. S., *Auditory Scene Analysis*, Cambridge, MA: MIT Press, 1990.

[12] Williams, S., "Perceptual Principles in Sound Grouping," in *Auditory Display: Sonification, Audification, and Auditory Interfaces*, G. Kramer, (ed.), Reading, MA: Addison-Wesley, 1994.

[13] Xenakis, I., *Formalized Music*, Bloomington, IN: Indiana University Press, 1971, Originally published as *Musique formelles*, by the Paris La Revue Musicale, 1963.

[14] Roads, C., *Microsound*, Cambridge, MA: MIT Press, 2002.

[15] Mauney, L. M., and B. N. Walker, "Individual Differences and the Field of Auditory Display: Past Research, a Present Study, and an Agenda for the Future," *Proceedings of the 13th International Conference on Auditory Display*, 2007, pp. 386–390.

[16] Walker, B. N., "Magnitude Estimation of Conceptual Data Dimensions for Use in Sonification," *Journal of Experimental Psychology: Applied*, Vol. 8, No. 4, 2002, pp. 211–221.

[17] Handel, S., *Listening: An Introduction to the Perception of Auditory Events*, Cambridge, MA: MIT Press, 1989.

[18] Steinberg, J. C., and W. B. Snow, "Auditory Perspective—Physical Factors," *Electrical Engineering*, January 1934, pp. 12–17.

[19] Cutting, J. E., "Auditory and Linguistic Processes in Speech Perception: Inferences from Six Fusions in Dichotic Listening," *Psychological Review*, Vol. 83, No. 2, March 1976, pp. 114–140.

[20] Kramer, G., "An Introduction to Auditory Display," in *Auditory Display: Sonification, Audification, and Auditory Interfaces*, G. Kramer, (ed.), Reading, MA: Addison Wesley, 1994.

[21] De Campo, A., "An Interdisciplinary Approach to an Environment for Sonification of Scientific Data," Ph.D Thesis, Institute of Electronic Music and Acoustics—IEM, University for Music and Dramatic Arts Graz, 2003.

[22] Kramer, G., et al., *Sonification Report: Status of the Field and Research Agenda*, National Science Foundation, February 1999.

[23] Childs, E. P., and S. Tomic, U.S. Patent 7135635—"System and Method for Musical Sonification of Data Parameters in a Data Stream," issued November 4, 2006, http://www.patentstorm.us/patents/7135635.html.

[24] Pereverzev, S. V., et al., "Quantum Oscillations between Two Weakly Coupled Reservoirs of Superfluid 3He," *Nature*, Vol. 388, July 31, 1997, pp. 449–451.

[25] Ballora, M., et al., "Heart Rate Sonification: A New Approach to Medical Diagnosis," *LEONARDO*, Vol. 37, February 2004, pp. 41–46.

[26] Baier, G., T. Hermann, and U. Stephani, "Multi-Channel Sonification of Human EEG," *Proceedings of the 13th International Conference on Auditory Display*, 2007, pp. 491–496.

[27] Hong, S. L., J. W. Bodfish, and K. M. Newell, "Power-Law Scaling for Macroscopic Entropy and Microscopic Complexity: Evidence from Human Movement and Posture," *CHAOS*, Vol. 16, 2006.

[28] Wallis, I., et al., "Real-Time Sonification of Movement for an Immersive Stroke Rehabilitation Environment," *Proceedings of the 13th International Conference on Auditory Display*, 2007, pp. 497–503.

[29] Hall, D., B. Hellar, and M. McNeese, "Rethinking the Data Overload Problem: Closing the Gap Between Situation Assessment and Decision Making," *Proceedings of the National Symposium on Sensor Data Fusion*, June 2007.

Chapter 8

Adapting IPB for Human Terrain Understanding: Informational Preparation of the Engagement Space[1]

"If you know the enemy and yourself, you need not fear the result of a hundred battles. If you know yourself and not the enemy, for every victory gained you will also suffer a defeat. If you know neither yourself nor the enemy, you will succumb in every battle."

—Sun Tzu, *The Art of War,* c. 400 B.C. [1]

War is no longer made up of set-piece battles between huge armies confronting each other with tanks and airplanes; warfare has become asymmetric, hybrid, and increasingly adaptive. Planning for operations whether during time of war or peace must reflect the dynamic nature of the threats facing a nation, whether from crime, terrorism, disease, or extreme weather. Planning and modeling of operational entities with regard to terrain have largely been the purview of the military operations and intelligence planners. Unfortunately, many of the planning tools and processes currently in use by military planners have neither kept pace with the rapid changes in the threat landscape nor leveraged the evolutionary advances in technology and social networking science.

This chapter was coauthored by Col. Jake Graham, USMC (Ret), College of Information Sciences and Technology, The Pennsylvania State University, University Park, Pennsylvania.

Planning for the employment and execution of military forces is an inherent responsibility of the commander at all echelons of command, across the full range of military operations: in the air, on land, at sea, and in space. The military decision-making process (MDMP) is the current doctrinal framework for decision-making and planning for tactical operations. Originated by the United States Army, adapted by the United State Marine Corps and used in varying forms by the Air Force and Navy, MDMP "is a planning tool that establishes techniques for analyzing a mission, developing, analyzing, and comparing courses of action against criteria of success and each other, selecting the optimum course of action, and producing a plan or order" [2]. MDMP has evolved over time and through war, arriving at its present state as a framework of standardized processes, procedures, and products through which military planners inform the commander to address the problems of war-making and peacekeeping in every clime and place. Integral to the MDMP is a visual-analytic process called Intelligence Preparation of the Battlefield (IPB). Like the military decision-making process itself, IPB seeks to inform the decision-maker using a structured framework of standardized processes to define the battlefield environment with regard to the operational imperatives of the various war-fighting functions. Unfortunately, like its parent analytic process, traditional forms of IPB lack the richness required to prepare tactical commanders for the uncertainty and fluidity of the human terrain.

This chapter explores the potential to adapt the traditional IPB process for human terrain understanding to address key aspects of sociocultural dynamics in the engagement space, as well as the likely pitfalls of the use of IPB to address and depict the human landscape. Such an enhanced IPB process could be utilized by military planners, crisis managers, business planners, or operational personnel in charge of nongovernmental organizations for disaster relief.

8.1 INTRODUCTION

Intelligence Preparation of the Battlefield (IPB) is a systematic, time-tested methodology of analysis for the study and understanding of the effects of battlefield conditions on enemy and friendly forces within a unit's operation and interest area. Army Field Manual (FM) 34–130, released in 1994, is the foundational document that describes the IPB process [3]. Integral to operational planning, the four-step IPB process supports military decision-making and plays a vital role in the development and evaluation of military plans at all levels of engagement. IPB is continuous and moderately adaptive; as battlefield conditions evolve, IPB is updated to keep the commander apprised of battlefield dynamics. IPB is structured to provide a solid framework for analysis across all branches and mission functions to identify essential information and intelligence requirements, shape collections, aid targeting, and inform command decisions [3, pp. 1–6] IPB is visual—one of its key strengths is in the use of layered graphics, depicting everything from weather to order of battle; IPB templates are easily adaptable and scalable to level of effort and desired understanding.

IPB is tailored to the unit's combat mission, that is, air defense, artillery, aviation, counterintelligence, electronic warfare, engineer, intelligence, chemical-biological-radiological-nuclear-explosive (CBRNE), signal, special operations, and rear and combat service support. In defining the battlefield environment, the analyst identifies the significant characteristics of the battlefield which may influence operations, friendly and enemy, according to the operational characteristics of the given unit and likely enemy actions, reactions, and capabilities.

That IPB continues to be an effective planning tool after more than 30 years of service is telling of its usefulness and adaptability within the military decision-making process. Evolved and adapted across generations of warfare and across each of the military services, IPB has been adapted to the operational focus of its users. Following the May 2000 release of Joint Publication 2–01, *Joint Tactics, Techniques and Procedures for Joint Intelligence Preparation of the Battlespace,* the U.S. Air Force released Air Force Pamphlet 14–118, June 2001, adapting and expanding the IPB process from a ground-centric model to one that includes military aerospace power and hence changing "battlefield" to "battle-space" and including the elements of both friendly and enemy space capabilities, theater ballistic missile defense, and force protection. The latest and arguably most significant transformation of the IPB process was revealed in Joint Publication 2–01.3, *Intelligence Preparation of the Operational Environment*, June 2009, establishing the standard for joint intelligence preparation of the operational environment (JIPOE), but primarily focused at the joint force level [4].

Agencies outside of the Department of Defense (DoD) have also recognized the utility and flexibility of the IPB process. The National Drug Enforcement Agency, for example, effectively introduced the tenets of IPB into its counter-narcotic operations, improving sense-making and resource allocation decisions within the counterdrug task force [5]. The National Geospatial Intelligence Agency (NGA) utilizes its own version of IPB that it calls *Geospatial Intelligence (GEOINT) Preparation of the Environment* [6]. The GEOINT Preparation of the Environment defines geographic boundaries and environmental influences on physical and cultural factors—but still the focus remains largely on the physical dimension, where planning considerations of geography, weather, climate, sea state, terrain, darkness, and time continue to dominate.

The most important output of IPB, or MDMP for that matter, is not the resulting plan, but the knowledge gained throughout the planning process. As Dwight D. Eisenhower aptly stated, "In preparing for battle I have always found that plans are useless, but planning is indispensable" [7]. Such plans often go out the door as soon as the first round goes down range. Sense-making and situational awareness of the likely effects of battlefield conditions on enemy and friendly forces are undoubtedly the most valuable by-product of IBP/MDMP; knowledge gained during planning is the best preparation for what von Clausewitz called the "fog of war," a term he penned to describe the commander's lack of clear understanding [8]. IPB seeks to develop that understanding and reduce the ambiguities and uncertainties of battle.

In a 2002 research study by the RAND Corporation for the U.S. Army, Jamison Medby and Russell Glenn produced *Street Smart: An Urban Approach to IPB*, which recommended improvements to the traditional IPB process to address the intricacies of urban combat [9]. Following that study, the Army produced FM 3–06, *Urban Operations,* and later, in December 2006, the Army and Marine Corps jointly released FM 3–24, *Counterinsurgency,* thus closing a 20-year doctrinal gap in addressing counterinsurgency operations [10]. In April 2009 the Department of the Army released Field Manual 3–24.2, *Tactics in Counterinsurgency*, establishing doctrine for tactical counterinsurgency (COIN) operations at the brigade level and below. The impact of FM 3–24 (and its companion FM3–24.2) goes beyond doctrine according to John Nagel, "The most important contribution of the manual is likely to be its role as a catalyst in the process of making the Army and Marine Corps more effective learning organizations..." The failures of post-Iraq war planning are testimony supporting Nagel's assertion that, "counterinsurgents . . . should strive to avoid imposing their ideals of normalcy on a foreign cultural problem" [11]. The emphasis on socio-cultural factors outlined in FM 3–24 and FM 3–24.2 should be formalized into the next iteration of FM 34–130.

8.1.1 Intelligence Preparation of Battlefield: The Traditional (Surface-Centric) Approach

"When I took a decision or adopted an alternative, it was after studying every relevant—and many an irrelevant—factor. Geography, tribal structure, religion, social customs, language, appetites, standards—all were at my finger-ends. The enemy I knew almost like my own side."

—T. E. Lawrence (*Lawrence of Arabia*), 1933 [3, p. 2–1]

The IPB process is about planning; it aids in predictions regarding the future state or actions of a target individual or group. Without defining the battlefield, the only option would be to make assumptions, which would lead to uninformed decisions. The benefit of defining the battlefield environment is that individuals involved have the information necessary to understand the environment they are dealing with. While T. E. Lawrence recognized the importance of understanding the socio-cultural elements of the enemy and the populous, its significance seems to have been lost by the drafters of FM 34–130. Traditional IPB fails to break out sociocultural considerations as a stand-alone planning/analysis step but chooses to fold human terrain considerations within step 2 of the process. Such factors as politics, civilian press, local population, and demographics are misplaced within this step while the characteristics of physical infrastructure and geography (including terrain and weather) are predominating.

The IPB process is effective, both as a stand-alone analytic process and as an integral part of MDMP; however, IPB's primary focus on the physical domain highlights its inherent weakness with regard to sociocultural understanding. In order to keep pace with the ever-expanding range of asymmetric threats facing our nation, whether from extreme weather events or terrorism, hybrid wars or Hurricane Katrina, the planning process must be equally considerate of the inhabitants occupying and influencing the engagement space. From relief operations to high intensity conflict, an adaptive IPB-like analysis process that accounts for the human element is essential.

How might the best of the traditional IPB processes be adapted for human terrain understanding as informational preparation of the engagement space? The following narrative provides a slightly more in-depth description of traditional IPB processes as described in current U.S. Army doctrine FM 34–130 followed by a brief discussion on how those processes have been adapted for counterinsurgency operations in FM 3–24 and FM 3–24.2. Additionally, it examines Chairman Joint Chiefs of Staff publication JP 2–01.3 Intelligence Preparation of the Operational Environment and offers discussion on the integration of JIPOE into service doctrine and practice regarding the human terrain.

8.2 THE TRADITIONAL (SURFACE-CENTRIC) IPB PROCESS

The four-phase IPB process aids in predictions regarding the future state or actions of a target individual or group—largely with regard to the physical domain (principally terrain and weather) on friendly and enemy actions. IPB is not battle planning by itself; it does not attempt to define tactical measures, but it informs battle planning and seeks to account for the effects of the physical domain on tactical operations. The benefit of defining the battlespace environment is that individuals involved have the information necessary to understand the environment with which they are dealing; ultimately IPB equates to sense-making.

Field Manual 34–130 describes the four steps of the IPB process[2] as:

1. Define the battlefield environment.

[2] Note that the four-step IPB process is analogous to the well-known OODA (observe, orient, decide, act) decision cycle introduced by Boyd.

2. Describe the battlefield effects.
3. Evaluate the threat.
4. Determine threat courses of action.

The process described next (and summarized in Table 8.1) represents a surface-centered approach to the IPB process that largely reflects the influence of its originator, the U.S. Army and, in a similar fashion, the U.S. Marine Corps. Given the ground-centric nature of the Army and Marine forces, this focus makes sense, and while the U.S. Army and U.S. Marine Corps each possess a considerable air arm, the center of gravity of each force resides with the infantry. The U.S. Air Force has expanded IPB from battlefield to battlespace, thus shifting from a geographic focus to more of a functional focus tailored to aerospace operations [12].

8.2.1 IPB Phase I

During Phase I, analysts will *Define the Battlespace Environment* [3, pp. 2–2 to 2–6]. This initial step sets the conditions for the entire process by focusing on the area in which operations are to be conducted. It is typically necessary to have some type of vision, goal, or mission to orient this initial stage. There are six steps that should be completed in Phase I prior to moving forward in the IPB process. First, the relevant and important characteristics of the environment must be identified. These characteristics will be dependent on the given mission and could range from location and activities of enemy long-range fire support (during war) to the availability of emergency medical services (during relief operations). Second, it is necessary to identify the limits of the area of operation (AO) and battlespace. Third, it is necessary to establish the limits of the area of interest (AI). The AO is

Table 8.1

Summary of IPB Process

Analysis Phase	*Key Question*	*Anticipated Outcome*
I. Define battlespace	What out there matters most?	Inventory of mission-relevant battlespace conditions
II. Describe battlespace effects	How does it impact us?	Understanding of what battlespace conditions allow
III. Evaluate the adversary	What can they do to us?	Assessment of sources of strength, capabilities, and vulnerabilities
IV. Determine adversary courses of action (COA)	What will they do?	Full set of prioritized and evaluated plans

the space in which authority has been provided to conduct operations, while the AI is a much larger space, typically utilized for the collection of the data necessary for planning.

Fourth, the necessary and appropriate level of detail must be established. This is essential, as only the information required to make decisions needs to be present. Gathering unnecessary information may simply add to the time required for data collection and expend resources without any additional benefits. When time and money are limited, one does not have the luxury of superfluous data collection. Therefore, the fifth step is an evaluation of the currently available and previously collected data. This assessment must be conducted to identify the information shortfalls. This is a critical step as it provides knowledge of the current intelligence gaps, which aids in knowing what information needs to be collected as well as some potential limitations and constraints for both analysis and decision making. This step should also address any assumption made earlier in the process. It may not be possible to collect all of the missing information in the designated timeframe, and it is important to know that. Understanding the capabilities of data collection and which data is of most relevance assists with the sixth and final step required to define the battlefield environment, which is to collect the data necessary to fill the knowledge gaps. Information gaps become priority information requirements (PIR) or information requirements (IR) and, where possible, are filled before the first round goes down range.

Phase I products include:

- Preliminary priority intelligence requirements (PIR) delineating the scope and detail required for the mission being planned, which evolve as the IPB process develops;
- The identification of significant battlefield characteristics affecting the commander's mission;
- The identification of intelligence gaps and priorities, which also evolve as the IPB process develops;
- An initial set of intelligence collection and production requirements that support further IPB analysis and the commander's mission.

8.2.2 IPB Phase II

In Phase II, analysts *Define the Battlespaces Effects* [3, pp. 2–7 to 2–28] using Phase I for focus. The characteristics of the battlefield impose constraints and opportunities to both enemy and friendly forces. This step seeks to identify geographic/environmental influences affecting probable enemy and friendly courses of action (COA). The analysis of the battlefield environment conducted in this step concentrates largely on the "military aspects of the terrain, and weather" [3].

The military aspects of terrain are defined as:

- Observation and fields of fire;
- Concealment and cover;
- Obstacles;
- Key terrain;.
- Avenues of approach.

The military aspects of weather are broadly defined as:

- Visibility;
- Winds;
- Precipitation;
- Cloud cover;
- Temperature and humidity.

Other factors that must be considered under this category include the effects of weather on weapons systems, subsystems, and personnel such as target acquisition systems, ballistic trajectories, mobility, and personnel performance.

"This phase considers general capabilities of each force until courses of action are developed in later steps of the IPB process" [3, p. 1–2]. Phase II draws attention to those military options that may be available to friendly forces and to those options that can be taken away from enemy forces. Characteristics of terrain considered in this step include the general aspects of geography, weather, infrastructure, and "other" characteristics of the battlefield. "Other" characteristics of the battlefield may include elements such as infrastructure, local population, and demographics. In assessing the vulnerabilities of key targets (both friendly and enemy), intelligence must be integrated with operational data to ensure that unit/system-specific mission considerations are addressed. This step will utilize engineer support when available to assess the terrain, but may be conducted manually through map analysis. Other specialized products address such factors as:

- Cross-country mobility;
- Transportation systems (road and bridge information);
- Vegetation type and distribution;
- Surface drainage and configuration;
- Surface materials (soils);
- Ground water;
- Obstacles.

Terrain analysis is a continuous process; thus, changes in the battlefield environment as a result of seasonal changes, weather patterns, or other battlefield effects, such as urban rubble or disruptions in the line of communications due to enemy or friendly obstacles, must be considered likely.

In the same way that weather and terrain are evaluated with respect to military aspects of operations, so, too, must "other" characteristics of the battlefield be evaluated. This catch-all category includes the balance of variables not already considered under terrain and weather. Other characteristics of the battlefield vary with circumstance and therefore a comprehensive list is not feasible. However, a partial list might include:

- Logistics infrastructure:
    - Land use patterns.
    - Sources of potable water.
    - Bulk fuel storage and transport systems.
    - Canals and waterways, with associated control facilities such as locks.
    - Communication systems.
    - Transportation means and systems, including road and rail networks, trans-loading facilities, and airfields.
    - Natural resources.
    - Industries and technologies.
    - Power production facilities.
    - Chemical and nuclear facilities.

- Population demographics, such as:
  - Living conditions.
  - Cultural distinctions.
  - Religious beliefs.
  - Political grievances.
  - Political affiliation.
  - Education levels.
- Economics.
- Politics:
  - Local.
  - Regional.
  - International (government systems, treaties, agreements, and legal restrictions; includes unofficial systems such as gangs).

The representation of information in a graphical form comes to the forefront in Phase II as analysts explore the impact that the battlefield may have on the capabilities of friendly and threat forces against most probable and most dangerous courses of action (COA). The information produced during the evaluation of the battlefield environment is integrated into a single graphic product called the *obstacle overlay*. By doing so, one is no longer looking at the factors of the environment individually. Instead, this combination allows a more holistic view of how the environment can impact COAs for operations (e.g., effects of "trafficability"—how terrain, roads, and vegetation affect the ability to move a group of vehicles from one location to another [13–15]. When evaluating the battlefield, it is important to do so not only by factors that compose it (terrain and weather) but from the perspective of the friendly or threat forces and the weapons systems, tactics, techniques, and procedures (TTPs) that they are anticipated to employ. It is necessary to focus on the perspective of the acting force (enemy or friendly) in order to allow for a more accurate determination of COAs.

Products developed during Phase II may include, but are not limited to:

- Population status overlay;
- Overlays that depict the military aspects and effects of terrain;
- Weather analysis matrix;
- Integrated products such as modified combined obstacle overlays (MCOOs).

The Joint Publication 2–01, *Intelligence Preparation of the Operational Environment* [16], describes the concept of a holistic view of the environment (see Figure 8.1). In that concept, an effort is made to link aspects of the physical terrain or domain (e.g., land, maritime, air, and space domains), with cyberspace and an information environment with system-level views of social, economic, military, and political aspects. Figure 8.1 indicates awareness of the need for understanding sociocultural factors and the human landscape. However, at this point the maturity of the information representation, modeling, and analysis processes is less than that of the traditional aspects of IPB.

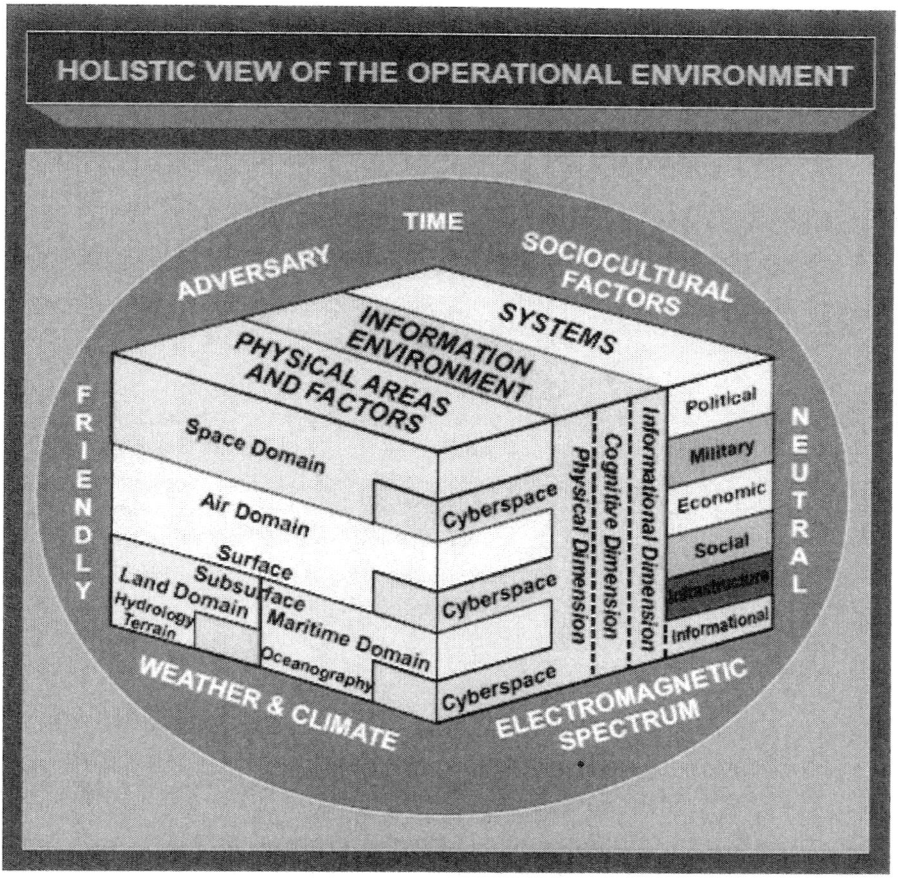

Figure 8.1 Holistic view of the operational environment.

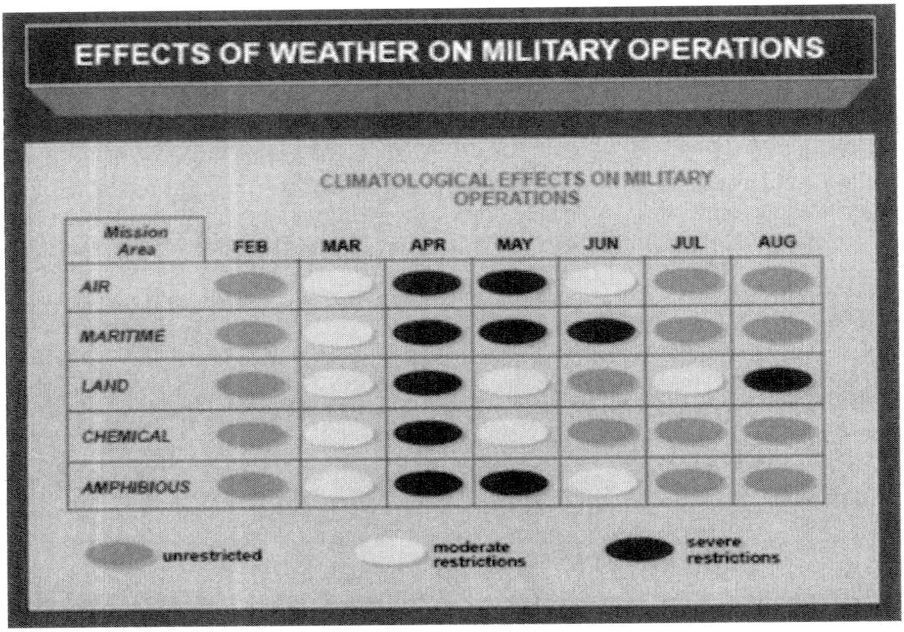

Figure 8.2 Example weather effects matrix [17].

Various tools and information presentation concepts have been developed for IPB. An example of a weather effects matrix is shown in Figure 8.2, which shows the climatology effects on different mission areas as a function of the time of year. An example of a modified combined obstacle overlay is shown in Figure 8.3. This shows conceptually how to combine different views of factors such as vegetation and surface drainage to create a combined obstacle view of the terrain for an area of interest.

8.2.3 IPB Phase III

The enemy is the focus of this phase of the IBP process as analysts *Evaluate the Threat* [3, pp. 2–29 to 2–38]. The essence of this phase is to "know thy enemy." In other words, this phase develops an understanding of the threats, capabilities, strengths, and weaknesses of your opponent. In this phase, it is necessary to determine the guiding principles and capabilities of the threat forces, as well as the tactics, techniques, and procedures (TTP) supported. During this phase, and

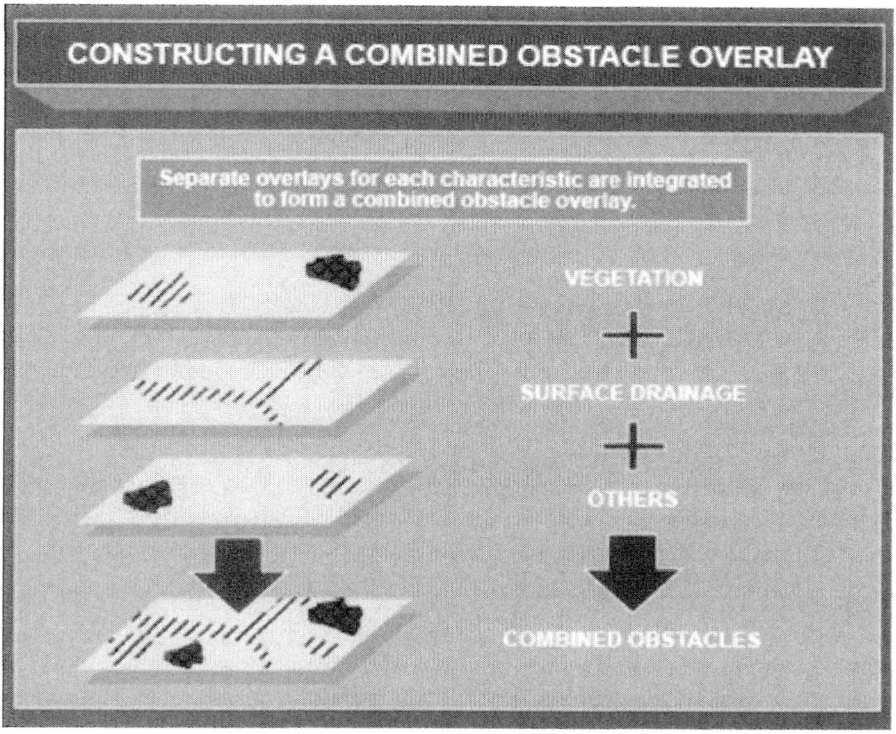

Figure 8.3 Example of modified combined obstacle overlay [18].

through the previous phases of analysis, a model is developed that demonstrates the way that the threat might initiate actions and react to friendly operations in a given situation based on the way their operations have been executed in the past. Effectively accomplishing this step requires a thorough examination of enemy doctrine, tactics, technology, weapons systems, and command and control as well as their ability to execute a given mission based on strengths and weaknesses, capabilities, tendencies, and past operational success or failure.

This phase contains two primary steps. The first step is to *create or update the threat models.* In order to do so, there are three things that need to be accomplished. First, graphical depictions (doctrinal templates) must be developed from known threat patterns. In other words, one must answer the question "How would this threat behave during ideal conditions?" and present the answer pictorially. This effort should be an ongoing process; the models should be constantly updated, refined, and evaluated as new information becomes available.

The model in itself is composed of three parts. The doctrinal template is the first part and is the part that has been previously discussed; it displays the target typical TTPs when there are no environmental constraints (ideal conditions). They are also developed to meet the needs of their creator. Therefore, a doctrinal template may not be as useful if shared with another group, say, one that has a different operational focus, or for another mission; they may not be as useful or relevant. For example, doctrinal threat templates developed for an armor unit may have limited utility for aviation forces.

The second part of developing the threat model is a description of tactics and options, which is essentially text that explains the doctrinal template (see, for example, Figure 8.4). This description should detail the options available to the threat force as well as their preferred route of approach. It should also be written from two sides: (1) the success of the mission, and (2) the failure of the mission. In other words, assuming successful execution, the enemy force is likely to prosecute operations according to this model. Much of this is inferred from past enemy actions. Conversely, if the enemy is not successful in the execution of this phase of operations, their likely response will be thus. This analysis also reflects knowledge of prior enemy operations or may be derived from an understanding of published

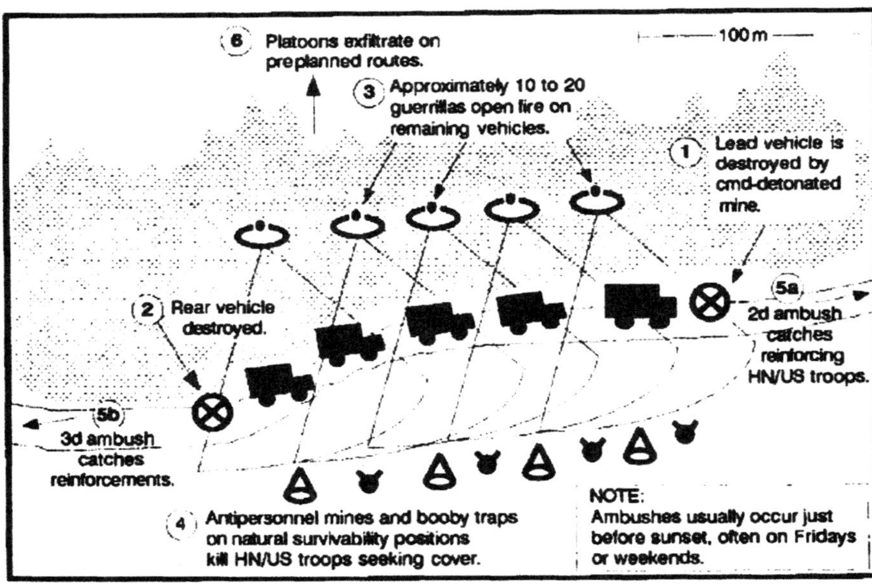

Figure 8.4 Doctrinal template depicts typified enemy tactics [3].

threat doctrine. In doing so, the doctrinal description will contain the alternative routes that the threat force is likely to follow should the mission not progress as planned.

The final part of the threat model is the identification of type high-value targets (HVTs). An example is provided in Figure 8.5. An HVT is defined as [17, p. xxiv], "A target the enemy commander requires for the successful completion of the mission. The loss of high value targets would be *expected to seriously degrade important enemy functions* throughout the friendly commander's area of interest." "Type" refers to the particular battlefield operating system (BOS) being modeled. Examples of BOS include aviation, maneuver, fire support, air defense, command and control, intelligence, mobility and survivability, and combat service support. This information is derived from the data gathered in the previous phases that provided intelligence regarding the threat, as well as associated doctrinal templates and their descriptions. It is essential to know which of these assets are crucial to the execution of the threat-force operation as well as the decision-making involved. The important thing to establish in this step is how the threat forces will respond to a situation in which their desired asset has been made unavailable to them.

The second and final step of Phase III is to *identify the capabilities of the threat.* Threat capabilities can be defined as broad courses of action and support operations available to the threat force that, if successfully executed, can influence

Description:

- Movement begins as early as 1.5 hours after order.
- Movement along previously rehearsed routes.
- Supported by well-planned and coordinated FS.
- Lead elements and supporting defenses fix friendly forces with fires.
- Main body seeks flanks of friendly forces.

Failure Options:

- Hasty defense.
- Fix friendly forces for counterattack by reserve of higher HQ.

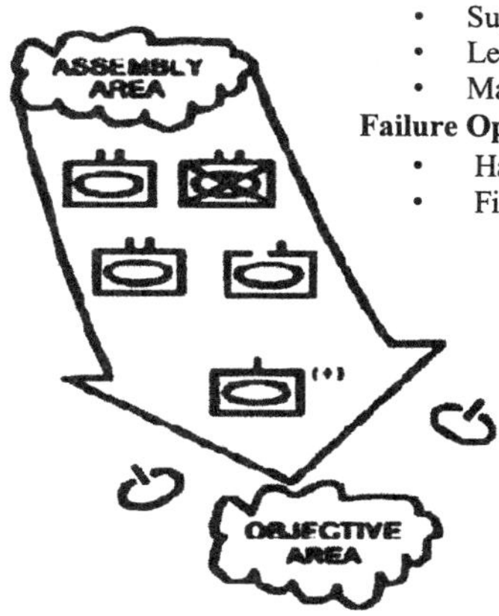

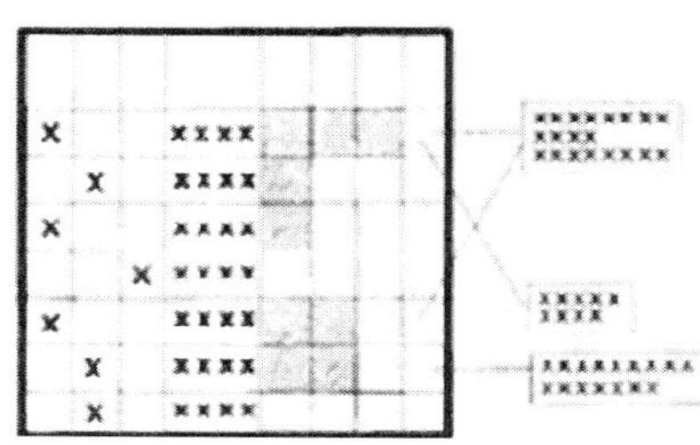

Figure 8.5 Depiction and description of high value targets (HTVs) [17].

the accomplishment of the friendly mission. Threat capabilities are often represented by broad statements such as:

- "The enemy has the capability to conduct an amphibious assault of 4 reinforced divisions with up to 800 sorties of transport helicopters, supported by naval gunfire and 1,200 daily sorties of fixed-wing aircraft."
- "The threat force can disrupt convoy operations on the main LOCs using VBIED and remote-detonated IEDs."
- "The enemy has the ability to establish rudimentary air defense warning systems by enlisting and forming the local population into an ad hoc observation force using mobile communication devices."

Considering conventional operations, there are four tactical COAs that are generally available to military forces:

1. Attack;
2. Defend;
3. Reinforce;
4. Conduct a retrograde.

These broad COAs can be further divided into a host of more specific COAs. For example, an attack may be defined by tactical formation (envelopment, penetration, and so forth) or by mode, air, indirect fire, ground, or amphibious assault.

In this step it is also necessary to identify the threats' potential operations and COAs that would hinder the ability of friendly forces being able to successfully accomplish its mission. The threat model should encompass the following six elements:

1. Standard graphic control measures, such as boundaries;
2. A description of typical tasks for subordinate units;
3. An evaluation of how well the threat force is trained on the task;
4. Employment considerations;

5. A discussion of typical contingencies, sequels, failure options, and wildcard variations;
6. An evaluation of the threat's strengths, weaknesses, and vulnerabilities, including an evaluation of typical HVTs.

Additional intelligence products that must be analyzed during this phase include the threat-force Order of Battle (OB) files. The exploitation of threat order OB information is crucial to the understanding of the threat forces' operational capabilities, tendencies, and weaknesses. OB files include:

- Composition;
- Disposition;
- Strength;
- TTPs (including habitual operating areas for unconventional warfare (UW) forces, gangs, insurgencies);
- Training status;
- Logistics;
- Effectiveness;
- Electronic technical data;
- Miscellaneous data (personalities, pseudonyms, other).

Successful execution of Phase III develops and refines an understanding of how the enemy operates and which weapons system or infrastructure elements are most critical to potential enemy COAs. Thus, the ultimate goal is to identify in general terms what operational goals the adversary seeks to achieve and what TTPs he will employ to achieve it.

8.2.4 IPB Phase IV

During Phase IV the analyst must determine the adversary's likely courses of action [3, pp. 2–39 to 2–51]. This step integrates the results of the previous phases of the IPB process and translates them into meaningful and likely conclusions. This portion of the analysis should illuminate specific areas where focused intelligence collection will reveal which COA the enemy is most likely to execute.

The final phase of the IPB process revolves around the process of *predicting the course(s) of action* the adversary is likely to pursue. In other words, what will they do and how will the likely COA influence the friendly force's ability to complete its mission? It is important to note that this is an educated guess. Therefore, there is the possibility that the prediction will be incorrect and the threat forces will behave in an unanticipated way. However, the goal of this phase is to prevent that. There are five steps required to complete this phase. First and foremost, the friendly force must explore the potential threat objectives and their overall goals. These objectives should not only deal with the traditional terrain objectives, but should also include things like economic and political goals. There exists the possibility that the threat-force objectives will not interfere with the mission of the friendly forces. However, this assumption should not be made without an exploration into these details.

The second step is to uncover as many COAs available to the threat force as possible; even those COAs that might indirectly impact the friendly force mission should be explored. In order to do so, it is important to view the situation from the threat-force point of view. To be successful in this step, the analyst must understand the threat-force doctrine and its likely influence on enemy behavior as well as the threat-force decision-making processes. Thus, while evaluating a given threat-force COA, it is essential to establish the logic behind it—from the threat-force point of view. In determining the COAs that the threat force can take, it is necessary to identify any redundancy in COAs and those that are not within the threat-force capability. This is done by reviewing the impact of the battlefield effects against the threat-force intent, dispositions, assets, capabilities, and vulnerabilities.

In exploring COAs, there are five criteria that should be met:

1. *Suitability* refers to whether the given COA, if executed, will lead to the accomplishment of the threat-force mission.
2. *Feasibility* takes into account whether time, space, and other resources will allow the COA to be successful.
3. *Acceptability* speaks to whether or not the threat can afford the consequences that go along with executing the given COA.
4. *Uniqueness* goes back to the idea of redundancy. If a given COA is considerably similar to another COA, then these actions cannot be considered independent but simply as variations.
5. *Consistency* with doctrine compares the threat-force principles with a given COA to determine whether or not they match.

Developing a full list of a threat's COAs will allow friendly forces to focus on those COAs that are most likely to impact the completion of their mission.

The third step in this phase is to analyze and prioritize all available threat-force COAs. There is no way to know for certain which COA the threat forces will utilize. Therefore, it is necessary to analyze and prioritize the full set of COAs based on the likeliness that a given COA will be selected. This prioritization of threat COAs is an ongoing process that may change as conditions change (environmental, political, or military). Therefore, it may be necessary to have different versions of the prioritized COA list. FM 34–130 identifies six steps that go into prioritizing each COA:

1. Analyze each COA to identify its strengths and weaknesses, centers of gravity, and decisive points.
2. Evaluate how well each COA meets the criteria of suitability, feasibility, acceptability, and consistency with doctrine.
3. Evaluate how well each COA takes advantage of the battlefield environment. How does the battlefield encourage or discourage selection of each COA?
4. Compare each COA to the others and determine if the threat is more likely to prefer one over the others. Most forces will choose the COA that offers the greatest advantages while minimizing risk.
5. Consider the possibility that the threat may choose the second or third "best" COA while attempting a deception operation portraying acceptance of the "best" COA.
6. Analyze the threat's recent activity to determine if there are indications that one COA is already being adopted. Does the threat's current disposition favor one COA over others?

The fourth step of Phase IV is to detail each COA as much as possible. The order in which these COAs are detailed should follow the order of the prioritization that occurred in the previous step. In detailing these COA the what, where, when, why, and how questions should be answered. What type of operation does the COA represent? Where will the COA be executed? When could the COA begin? Why would the threat force execute this COA (what is the objective)? How will the threat force expend its resources?

There are three elements required to complete the details of each COA: a situation template (see Figure 8.6), a description of the COA and options, and a listing of associated HVTs. It is important to note that more detail will be required for COAs considering a defending threat than would be required for COAs considering an attacking threat.

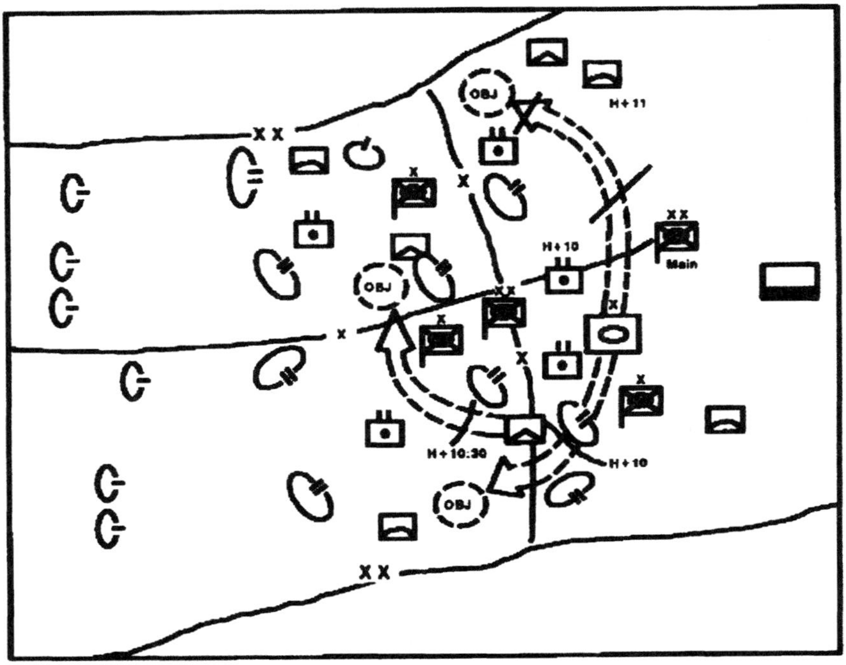

Figure 8.6 Situation template depicts likely threat-force COA [3].

The fifth step after identifying potential threat-force COAs is to identify intelligence/information collection requirements. Since friendly forces do not know without a doubt which COA the threat force will select, the correct identification of collection requirements is essential for the accurate determination of the threat-force COA. Developing an event template informs friendly forces of which COA the threat force is most likely to execute. It does so by indentifying the friendly force areas (areas of operation or influence) from which to collect information that will serve as indicators and warning that a particular COA is under way. The event matrix (see Table 8.2) is another support tool for intelligence gathering. It supports the event template by providing details on the type of activity expected in each named area of interest (NAI), the times the NAI is expected to be active, and its relationship to other events on the battlefield. A discussion of computer-automated templating methods is provided by Hall and Linn [19].

Table 8.2
The Event Matrix Supports the Event Template [3]

NAI No.	*No Earlier Than (Hours)*	*No Later Than*	*Indicator*
NAI 1	H-7	H-2	Engineer preparation of artillery positions
NAI 1	H-2	H-30 min	Artillery occupies firing positions
NAI 1	H-1	H-15 min	Artillery commences preparatory fires
NAI 2	H-2	H-1.5	Combat recon patrol conducts route recon
NAI 2	H-1.5	H-30 min	Rifle company (+) in march formation

8.3 IPB Is Continuous

It is worthy of noting that military IPB is not conducted by a single entity, say, at the headquarters level and then force-fed down to subordinate units. IPB is conducted at all levels of engagement and by each of the war-fighting functional entities, with each element accounting for the unique effects of the battlefield environment on their particular function and vice versa. "As the size of the unit increases, the level of detail required in the IPB effort increases significantly" [3, p. 1–1]. IPB tends to be a formal process at division/wing and higher echelons and less formal at battalion/squadron levels where planning is an expansion of support to the higher mission.

The IPB process aids in predictions regarding the future state or actions of a target individual or group, largely with regard to the physical domain (principally terrain and weather) on friendly and enemy actions. IPB is not battle planning and it does not attempt to define tactical measures, but it informs battle planning and seeks to account for the effects of the physical domain on tactical operations. Without defining the battlefield, the only option would be to make assumptions, which would lead to uninformed decisions. The benefit of defining the battlefield environment is that individuals involved have the information necessary to understand the environment they are dealing with; ultimately IPB equates to sense-making.

IPB makes extensive use of graphics, including annotated maps and multi-layered overlays. Additional intelligence products, when available, augment the IPB toolkit, including gridded photographs, microfilm, and detailed target analysis products. For the majority of IPB users, graphics are rendered manually, making it a time-consuming and tedious task. Efforts to automate the IPB process are currently being pursued by the various service components to reduce the manpower and time required to conduct the IPB process.

Systems like Zel Technologies' Automated Assistance with Intelligence Preparation of the Battle-space (A2IPB) are being developed for the U.S. Air Force and have the potential to allow the commander and his staff to war-game from within a software package. Friendly courses of actions are graphically portrayed against an adversary's most likely/most dangerous COA and played out electronically (see Figure 8.7). All relevant information is loaded into the same database, war-gaming is facilitated, data modification is easier, and the ability to play back a particular set of COAs multiple times is possible [17].

Whether automated IPB systems will replace current manpower intensive and time-consuming methodologies is yet to be determined, but with the advancements in computing speed, graphical interfaces, and data fusion, the possibilities are promising.

8.3.1 Counterinsurgency (COIN) and the New JIPOE

Joint Publication 1–02 defines *insurgency* as an organized movement aimed at the overthrow of a constituted government through the use of subversion and armed conflict [18]. While insurgency may rise from revolution, it is not a spontaneous phenomenon emerging in response to government action or inaction. It represents armed rebellion against a constituted authority designed to weaken the control and legitimacy of an established government or occupying power. *Counterinsurgency* combines military, paramilitary, political, economic, psychological, and civic actions taken by the constituted authority or occupying power to defeat insurgency [18, p. 109].

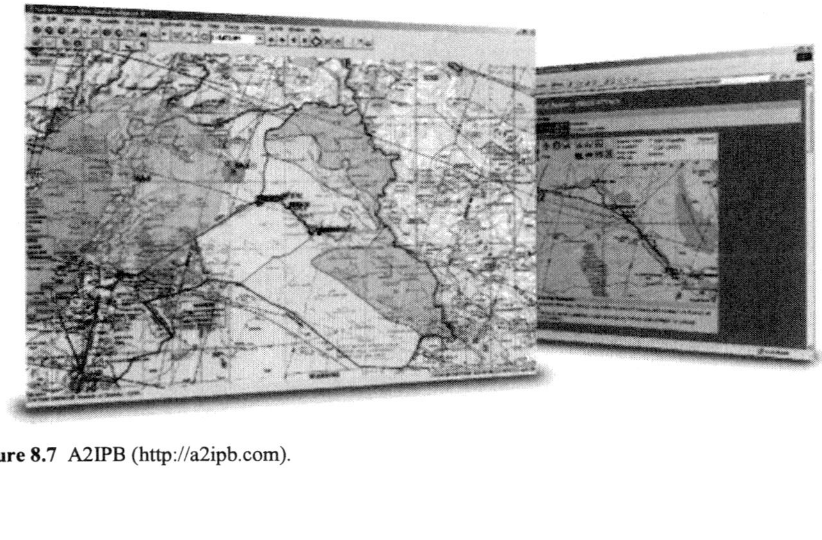

Figure 8.7 A2IPB (http://a2ipb.com).

The *Counterinsurgency Manual*, coauthored by the Army and Marine Corps, released in 2006, was the first major military publication written specifically to address the doctrinal gap between large-scale conventional warfare and modern-day insurgency operations. FM 34–2 evolves traditional military doctrine to address the challenges posed by a global asymmetric threat, one that is able to combine commercial-off-the-shelf products and social network theory with Web-based applications to challenge the technological wherewithal of America's military war machine.

8.3.2 IPB in COIN

IPB in COIN operations follows the broad traditional approach and methodology detailed above derived from FM 34–130; however, it places greater emphasis on the human element (what is referred to as "civil considerations"), especially people and leaders in the AO, than does IPB for conventional operations. IPB for COIN operations looks beyond the kinetic view of the battlefield to consider the less tangible issues of attitude, politics, family ties, tribal culture and ethnicity, religion, and an ever-expanding list of other human factors. IBP is largely a graphic process; unfortunately, the graphic representation of these influences presents many challenges, namely, how does one actually overlay attitude on a terrain map, and at what level do you chart (individual or group)?

The brief review of the FM 34–2 *Counterinsurgency Manual* reveals an analytic approach with an emphasis on the human dimension:

Step 1: Defining the Area of Operations (AO)

Defining the area of operations (AO) considers the commander's view of the physical geography with a special consideration of civil affairs, particularly human factors. It recognizes that lines of communications are not physically bounded by road or border but are more often defined tribally, economically, or culturally.

Likewise, the greater area of interest can be large relative to the AO, and therefore, the commander must account for various influences that affect the AO, such as:

- Family, tribal, ethnic, religious, or other links that go beyond the AO;
- Communication links to other regions;
- Economic links to other regions;
- Media influence on the local populace, U.S. public, and multinational partners;
- External financial, moral, and logistic support for the enemy.

Step 2: Describe the Effects of the Operational Environment

Describing the effects of the Operational Environment involves developing an understanding of the environmental factors within the operating area likely to impact the success of operations. Again, this step broadens the efforts of traditional IPB through an in-depth analysis of what FM 34–2 calls "civil considerations." The major elements of analysis during this step include:

- Civil considerations, with emphasis on the people, history, and host nation government in the AO;
- Terrain analysis (physical geography), with emphasis on complex, suburban, and urban terrain;
- Key infrastructure;
- Lines of communications;
- Weather.

Civil considerations addressed in this step seek to determine how the man-made infrastructure, civilian institutions, and attitudes and activities of the civilian leaders, populations, and organizations within an area of operations influence the conduct of military operations. Civil considerations comprise six characteristics, expressed in the mnemonic ASCOPE: areas, structures, capabilities, organizations, people, and events. The complexities of this analysis are highlighted by the drill-down analysis performed on the elements of ASCOPE as outlined next.

In order to evaluate the *people*, the following six sociocultural factors are analyzed:

1. Society;
2. Social structure;
3. Culture;
4. Language;
5. Power and authority;
6. Interests.

Social Structure

1. Groups:
   a. Racial
   b. Ethnic
   c. Tribes
   d. Religious
2. Groups are defined by relationships:
   a. Formal
   b. Informal
   c. Divisions and cleavages between
   d. Cross-cutting ties (religious alignments that cross other group boundaries)
3. Networks
4. Institutions
5. Organizations
   a. Types of organizations are further defined as:
      i. Communicating
      ii. Religious
      iii. Economic
      iv. Social
6. Roles and statuses
7. Social norms
   a. Some norms that may impact military operations include the following:
      i. The requirement for revenge if honor is lost.
      ii. Appropriate treatment of women and children.
      iii. Common courtesies, such as gift giving.
      iv. Local business practices, such as bribes and haggling.

Culture

Commanders should consider groups' attitudes regarding the following:

- Other groups
- Outsiders
- HN government
- United States
- U.S. military
- Globalization

Power and Authority

The formal political power system includes the following organizations:

- Central governments
- Local governments
- Political interest groups
- Political parties
- Unions
- Government agencies
- Regional and international political bodies

Authority:

- Rational-legal
- Charismatic
- Traditional

Step 3: Evaluate the Threat

Evaluating the threat analyzes the prominent characteristics of the adversary. This step also departs from previous IPB approaches where attempts to apply traditional order of battle factors and templates can produce oversimplified, misleading conclusions when applied to an insurgency threat. Commanders require knowledge of difficult-to-measure characteristics. These may include the following:

- Insurgent goals;
- Grievances that insurgents exploit;
- Means that insurgents use to generate support;
- Organization of insurgent forces;
- Accurate locations of key insurgent leaders.

Insurgency characteristics, such as objective, motivation, and means of generating popular support, are often the commander's most important intelligence requirements and also the most difficult to ascertain. Intelligence organizations fuse data and information into all-source intelligence products to support COIN operations. A comparison of insurgency characteristics and conventional order of battle factors is shown in Table 8.3.

Table 8.3

Insurgency Characteristics and Other Order of Battle Factors [10, p. 3–13]

Insurgency Characteristics	*Conventional Order of Battle Factors*
Insurgent objectives	Composition
Insurgent motivations	Disposition
Popular support or intolerance	Strength
Support activities, capabilities, and vulnerabilities	Tactics and operations
Information activities, capabilities, and vulnerabilities	Training
Political activities, capabilities, and vulnerabilities	Logistics
Violent activities, capabilities, and vulnerabilities	Operational effectiveness
Organization	Electronic technical data
Key leaders and personalities	Personalities
	Miscellaneous data
	Other factors

Analysis for COIN operations is very challenging, due in part to the:

- Need to understand perceptions and culture;
- Need to track hundreds or thousands of personalities;
- Local nature of insurgencies;
- Tendency of insurgencies to change over time.

Step 4: Determine the Threat Courses of Action

Determining the threat courses of action is intended to develop an understanding of the insurgent tactics, techniques, and procedures in order to develop counter-insurgent measures to defeat or neutralize them. Two levels of analysis are conducted in this step: The first is to determine the overall (operational) approach or combination of approaches likely to be undertaken by the insurgency force. The second level of analysis attempts to derive the level of action (tactical) methods to be utilized to execute the operational objectives. There are six approaches that insurgencies are likely to follow:

1. Conspiratorial;
2. Military-focused;
3. Urban;
4. Protracted popular war;
5. Identity focused;
6. Composite and coalition.

Complicating matters is the fact that insurgencies often adopt hybrid strategies that combine several approaches and/or transitions from one approach to another during the span of conflict. The following list highlights some of the key aspects of the varying approaches:

- Conspiratorial
    - Absence of overt violent or informational actions.
    - Large cadre relative to the number of combatants in the organization.
    - Small mass base or no mass base at all.

- Military-focused
    - Presence of leaders and combatants in the organization.
- Urban
    - Terrorist attacks in urban areas.
    - Infiltration and subversion of hot-nation government and security force in urban areas.
    - Organization composed of small, compartmentalized cells.
    - Cadre and mass base small relative to the number or combatants.
- Protracted popular war
    - A large mass base.
    - Overt violence.
    - Heavy use of informational and political activities.
    - Focus on building popular support for the insurgency.
- Identity-focused
    - Presence of a resistance movement.
    - Presence of an "us-and-them" gap between the government and one or more ethnic, tribal, or religious groups.
    - Large mass base of passive and active supporters built around preexisting social networks.
    - Many auxiliaries.
    - Small cadre composed primarily of traditional authority figures.
    - Large number of part-time combatants.

Counterinsurgency is arguably the most complicated form of warfare presenting unique challenges to large-scale military forces whose policies, strategies, tactics, and technology acquisition programs have been forged to fight in a major regional conflict (MRC). Effectively addressing the counterinsurgency threat requires new analytic approaches. The adaptive nature of the threat requires an equally adaptive response. Large-scale warfare has historically required large-scale, detailed, and deliberate planning; small-scale warfare, such as counterinsurgency, requires rapid-response planning that considers the dynamic and evolutionary nature of technology, social networking, and the Web. Counterinsurgency responses must also take into account the evolving socio-behavioral characteristics of today's generation of Web users on both sides of the issue. This begins with an understanding of the motivations, methods, and modes of the digital native.

8.3.3 Joint Intelligence Preparation of the Operational Environment (JIPOE)

JIPOE was promulgated by the Chairman, Joint Chiefs of Staff, in Joint Publication 2–01.3. JIPOE is the analytical process used by joint-force planners "to construct intelligence assessments, estimates, and other intelligence products in support of the Joint Force Commander's (JFC) decision-making process" [4, p. xi]. JIPOE expands traditional military battlespace to include the total operational environment through a thorough examination of the physical areas and factors as well as the informational elements that comprise the joint force areas of operation, influence, and interest. The physical areas include those domains required to

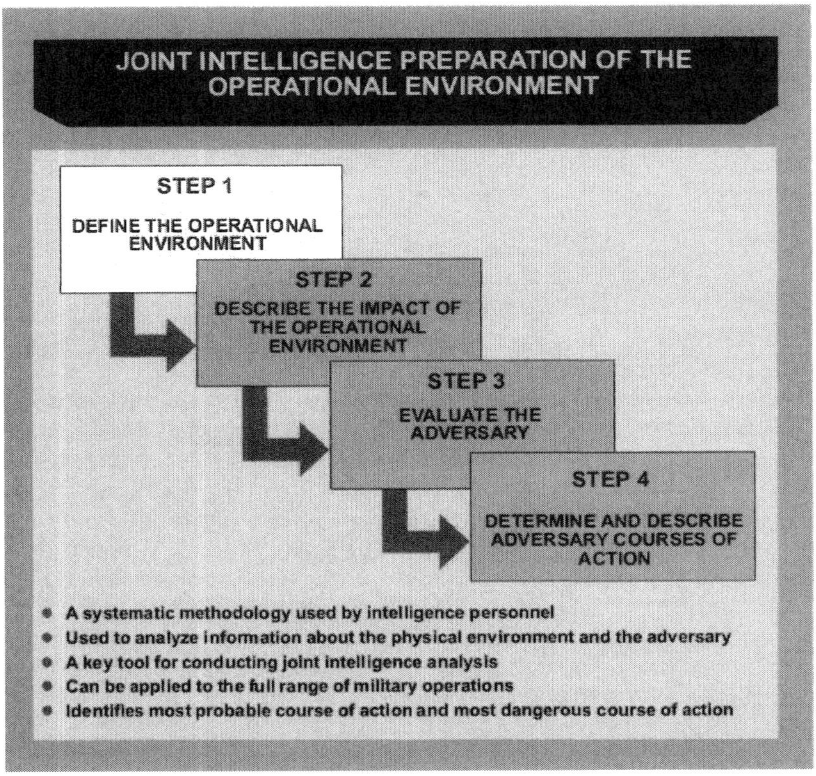

Figure 8.8 Joint Intelligence Preparation of the Operational Environment [20, p. I–17].

conduct operations on land, sea, air, and space; whereas "the information environment is the aggregate of individuals, organizations, and systems that collect, process, disseminate, or act on information" [4, p. xii]. The June 2009 revision of JP 2–01.3 expands and replaces Joint Publication 2–01.3 dated May 2000, and provides guidance on the use of JIPOE products, methods, and analysis to join forces with special consideration given to irregular warfare and stability operations. An overview of the Joint Preparation of the Operational Environment is shown in Figure 8.8. A brief summary of the four-step JIPOE process follows.

Step 1. Define the Operational Environment: identifies key aspects and significant characteristics relevant to the joint force mission. Seven elements define the operational environment:

1. Identify the joint force operating area.
2. Analyze the mission and the joint force commander's intent.
3. Determine significant characteristics of the operational area.
4. Establish the limits of the joint force's areas of interest.
5. Determine the level of detail required and feasible within the time available.
6. Determine intelligence and information gaps, shortfalls, and priorities.
7. Collect material and submit requests for information to support further analysis.

Step 2. Describe the Impact of the Operational Environment: seeks to evaluate the impact of the operational environment on friendly, adversary, and neutral military (or in the case of hybrid warfare operations—terrorist) capabilities and associated broad courses of action. Three elements comprise this step:

1. Develop a geospatial perspective of the operational environment.
2. Develop a systems perspective of the operational environment.
3. Describe the impact of the operational environment on adversary and friendly capabilities and broad courses of action.

Step 3. Evaluate the Adversary: identifies and evaluates the adversary's current situation, capabilities, limitations, centers of gravity, doctrine, tactics, techniques, and procedures (TTPs) as well as established patterns and trends of the adversary forces. Four elements make up this step of the JIPEO:

1. Create or update enemy models.
2. Determine the current adversary situation.
3. Identify adversary capabilities and vulnerabilities.
4. Identify adversary centers of gravity.

Step 4. Determine Adversary Courses of Action: translates knowledge gained in the first three steps into a holistic perspective in order to determine the adversary's probable intent and strategic objectives. Essential to this step is the identification of the range of possible adversary COAs in order to determine the most likely COA and the COA most dangerous to friendly forces. Five elements make up this step:

1. Identify the adversary's likely objectives and desired end state.
2. Identify the full set of adversary COAs.
3. Evaluate and prioritize each COA.
4. Develop each COA in the amount of detail that time allows.
5. Identify initial collection requirements.

A high-level comparison of the four-step JIPOE process to traditional IPB and IPB for COIN operations depicted next in Table 8.4 reveals a complementary set of processes; however, some key distinctions are worth noting, most notable are purpose, focus, and level of detail. "The purpose of JIPOE is to support the JFC by determining the adversary's probable intent and most likely COA for countering the overall friendly joint mission, whereas IPB is specifically designed to support the individual operations of the component commands" [4, p. I–4]. JIPOE conducts analysis at the macro level focusing on the adversary's strategic initiatives, whereas IPB requires a more finite examination of friendly and adversary elements at the tactical and operational levels.

Table 8.4

Comparison of Intelligence Analysis Processes

Traditional IPB	*IPB-COIN*	*JIPOE*
1. Define the battlespace	1. Define the area of operations (AO)	1. Define the operational environment
2. Describe the battlespace effects	2. Describe the effects of the operational environment	2. Describe the impact of the operational environment
3. Evaluate the adversary	3. Evaluate the threat	3. Evaluate the adversary
4. Determine adversary courses of action	4. Determine the threat's courses of action	4. Determine adversary's courses of action

Like its predecessor, JIPOE makes use of visual analytics in order to orient and inform decision-making. JIPOE presents a systems perspective of the operational environment to provide the JFC with a meaningful understanding of the connections, relationships, interactions, and effects of PMESII on the overall friendly mission as well as the effect of agency and coalition actions on JFC operations (see Figure 8.9). And like IPB, JIPOE is continuous with an aim at evaluating adversarial courses of action (COA) and, in particular, the adversary's most likely COA, as well as identifying the COA most dangerous to friendly forces. At the heart of this process is the time-tested center of gravity analysis, conducted to identify adversary centers of gravity in order to direct and focus intelligence collection and to achieve information superiority.

JIPOE is not intended to replace or supplant traditional IPB. Its holistic approach is intended to integrate analysis processes, joining geospatial intelligence

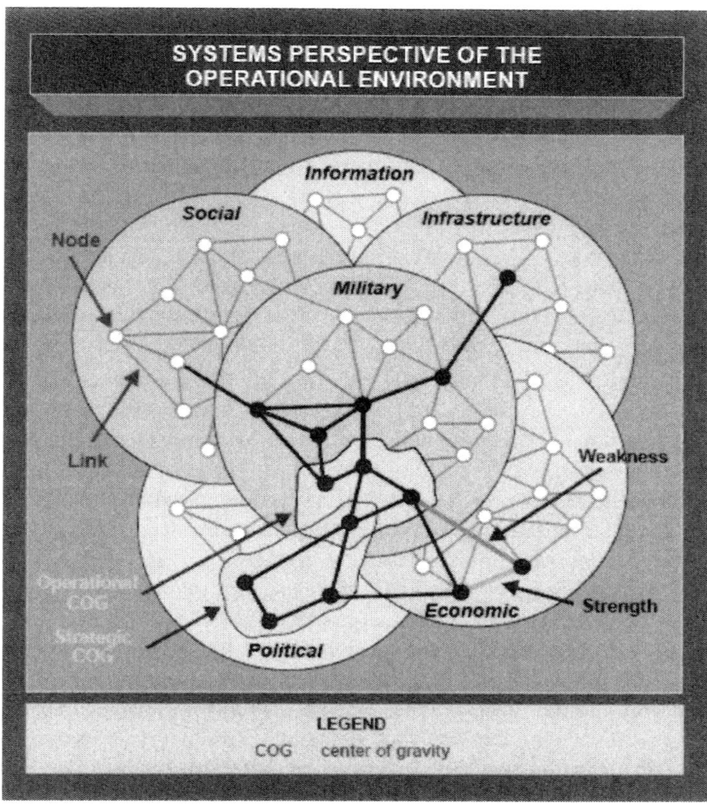

Figure 8.9 Systems perspective of the operational environment [4, p. II–45].

with force-level IPB from component and multinational forces and supporting agencies to gain "a synergistic integration of perspectives" [4, p. I–5]. JIPOE also complements evolving joint doctrine for effects-based operations (EBO), integrating the various elements of the political, military, economic, social, infrastructure, and informational (PMESII) within the operational environment [21].

8.4 ADAPTING IPB TO JIPOE FOR THE HUMAN TERRAIN

The *COIN Manual* and its complement, *Tactics in Counterinsurgency*, address the need to focus IPB-type activities on the human landscape as well as the physical landscape. As previously noted, however, a lag exists between the types of displays and models between traditional physical terrain-focused IPB and an enhanced human-terrain IPB. Thus, we argue that the IPB process, including sub-processes, support tools, displays, and analysis processes, will need to evolve to a human-terrain IPB process. JIPOE offers a framework to close the doctrinal gap; however, it does not solve the problem entirely. The elements of PMESII are appropriate for nation-state engagements where facilities, institutions, and other PMESII systems are tangible entities. Nonstate actors are asymmetric and networked lacking established and recognizable PMESII systems as noted by former Chief of Staff, U.S. Army, General Peter J. Schoomaker in his message to the force, *Serving a Nation at War* [21]. While the elements of PMESII are not perfect for each adversary, the proof is in the process, where "Planning will be iterative and collaborative rather than sequential and linear, more a framework for learning and action than a rigid template" [21, p. 16].

As previously discussed, the traditional IPB process focuses on developing an understanding of the potential conflict environment, seeking to define the battlefield environment, describe battlefield effects, evaluate the threats, and determine threat courses of action. In a conflict situation involving primarily physical weapons interacting with the physical enemy (i.e., tanks, aircraft, ships, and physical sensors), IPB necessarily focuses on the physical environment (weather, terrain), enemy physical systems, and courses of action and the interaction between physical entities (directed by human adversaries). While it is an oversimplification, the data and models for the traditional IPB approach involve primarily the physical world and human infrastructure and objects (e.g., transportation, human settlements, and communications). Tables 8.5 and 8.6 provide a summary of the types of data involved in this approach [22–46]. In essence, these are the types of information and displays that could be obtained from Google Earth or a GPS navigation device such as a TomTom [33] or

Table 8.5
Examples of Physical Terrain/Environment

Category of Data	*Types of Data*	*Example Data Sources*	*How Observed or Collected*	*References*
Terrain	Elevation data, slope data, digital elevation models (DEMS), digital terrain elevation model (DTED)	National Geophysical Data Center; National Geospatial Intelligence Agency, U.S. Geological Survey (USGS), Tele Atlas, NAVTEQ	Satellite data (e.g., optical, synthetic aperture radar); Ground-based human surveyors	[25, 26, 31, 32]
Geology and natural resources	Mineral deposits, oil deposits	USGS Earth Science Photographic Archive [27], IUGS TechTrask [28]	Observation of variations in magnetism, Geiger counters, aerial photography, airborne electromagnetic surveys, satellite-based spectroscopy, human prospecting	[27, 28]
Hy-drography	Surface water (oceans, rivers, lakes, streams, aqueducts), underwater supplies (embedded water) Bathymetry, flood planes and areas, hydrometeorology data	USGS National Hydrography Dataset, national integrated drought information system (www.drought.gov)	Sonar data, satellite data	[3, 29]
Weather	Surface features (highs, lows, fronts); precipitation; clouds and cloud cover; temperature contours; pressure contours, wind data	Accuweather *Otherweather* U.S. National Weather Service	Weather satellites Local human observers (e.g., *otherweather* crowdsourcing); Ground-based weather stations; monitoring moisture around cell phone towers [41]	[23, 24, 41]
Natural vegeta-tion	Plants and trees that are natural to an area (prior to agricultural cultivation)	United Nations Environmental Programme U.S. Environmental Protection Agency	Same as above, augmented by agriculture human agents, park forest personnel	[34, 35]

Table 8.6

Examples of Human Infrastructure and Objects

Category of Data	*Types of Data*	*Example Sources*	*How Observed or Collected*	*References*
Trans-portation	Roads, bridges, dams, rail lines, maritime ports and routes, airports, and airline routes	NAVTEQ; www.vesseltracker.com; Automatic Identification System (AIS)	AIS system, satellite images, ground-based radar	[31, 36]
Agriculture	Types of crops; location and types of animals and livestock; seasonal variations in crops	FAOSTAT (http://faostat.fao.org); U.S. Dept. of Agriculture, country-specific agriculture, National Oceanographic and Atmospheric Administration (NOAA)	Satellite images, ground-based human observers, aircraft observations	
Energy	Energy plants (nuclear plants, power plants), power lines	World Nuclear Association	Satellite images, ground-based human observers, aircraft observations, surveys, public reports; ground penetrating radar (for underground lines)	[37, 38]
Human settlements and facilities	Towns and cities, medical facilities, military facilities, industrial facilities, schools, places of worship	NAVTEQ		[31]
Commerce	Markets, banks, ATMs, payroll advance, money-exporting services (as in NYC for immigrants)	New York Stock Exchange; NASDAQ, Dow Jones; S&P 500	Financial reports; bank transactions; money exchange transactions	[42]
Communica-tions	Phone land lines, cell phone towers, coverage footprints, radio and TV stations, newspaper printing services	GSM world cell phone coverage [39]; Antennasearch.com	Planning data bases; physical surveys; aerial photography	[39, 40]

Table 8.7
Examples of Human Terrain Data

Category of Data	*Types of Data*	*Example Sources*	*How Observed or Collected*	*References*
Population	Population density	CIA *World Factbook*; U.S. Dept of State	Census surveys	[44–46]
Political divisions	Political boundaries (countries, states)	CIA *World Fact book*; United Nations (UN)	Self-reported by local governments	[44–46]
Economic conditions	Trading networks, currency, crop health, port activity	Gallup world poll; UN	In-country surveys	[43–46]
Ethnic composition		Gallup world poll; UN	In-country surveys	[44–46]

Magellan GPS [22] device or even cell phone-based applications. By contrast, Table 8.7 provides examples of information required to characterize the human landscape.

Extending IPB processes to the human landscape will require a focus on new types of data, new sources of information, new models, and new analysis techniques. While we do not have the solution to this problem, we do recommend a systematic investigation of the human-terrain IPB process, including an analysis of the utility of physical versus human information sources, types of representation models and prediction limitations, and new analysis methods for understanding how the human terrain affects situations and missions.

8.5 SUMMARY

Intelligence Preparation of the Battlefield (IPB) is a proven methodology for systematically assessing missions involving potential conflict with adversaries. The approach involves a four-step approach that: (1) defines the enemy environment, (2) describes battlefield effects, (3) evaluates potential threats, and (4) determines threat courses of action. The approach has been well documented in doctrinal field manuals and has been updated for counterinsurgency involving asymmetric operations. IPB has been extended from U.S. Army operations to U.S. Air Force, Marine, and Navy operations and has found applications for related areas such as crime prevention and counternarcotics. Changes in military focus (from an emphasis on the physical terrain and kinetic warfare to the human terrain and "soft" engagements) implies that, in order to be of continued use, the IPB process needs to be updated for the human terrain. This chapter has provided an overview of the traditional and evolving IPB process and recommendations for potential enhancements of IPB for current human-centered fusion environments.

The recent doctrinal implementation of JIPOE does not make obsolete the tenets or practice of IPB, but rather it serves as a top-level guidance to shape planning processes (including IPB) across the engagement space. As has been stated repeatedly, IPB remains valid in its approach and intent; however, much can be benefitted from a fresh look at IPB vis-à-vis JIPOE. The use of a systems perspective to define key interrelated adversarial systems and guide analysis is not new; the lexicon of planning acronyms grows with each generation of warfare. Whether using DIME, PMESII, ACOPE, or any number of other mnemonics to define the key elements of analysis, in the end it is the process of analysis and knowledge gained throughout that is most important, *not* the analysis tool de jour. When the U.S. Army and Marine Corps collaborated on FM 3–24, *Counterinsurgency* (COIN), and later on FM 3–24.2, *Tactics for Counterinsurgency*, they continued to cite the tried and true of the 1994 publication, FM 34–130, *Intelligence Preparation of the Battlefield*. FM 3–24, COIN utilizes the acronym, ASCOPE (areas, structures, capabilities, organizations, people, and events), to guide the analysis of civil considerations, with no mention of PMESII in this component-level 2006 document. The April 2009 release of FM 3–24.2 predating Joint Publication 2–01.3 by only a few months looks to close the doctrinal gap, if only by inches by making use of PMESII to partially define the operational and mission variables. By adding the elements of physical environment and time to PMESII, it appears that the gap between service and joint doctrine is closing, a step that may provide the impetus for refining IPB in the next release of FM 34–130.

References

[1] Sun Tzu, *The Art of War,* "Part III: Attack by Stratagem," Verse 18, Translation and critical notes by Lionel Giles, London: British Museum, 1910, p. 32.

[2] U.S. Army Field Manual, 5–0, Army Planning and Orders Production, final draft, Washington, D.C., Government Printing Office, 2003, p. 3–1.

[3] Field Manual 34–130, *Intelligence Preparation of the Battlefield*, Headquarters, Dept. of the Army, 1994, 1–1.

[4] Joint Publication 2–01.3, *Intelligence Preparation of the Operational Environment*, Office of the Chairman of the Joint Chiefs of Staff, June 16, 2009, p. iii.

[5] http://www.globalsecurity.org/military/library/report/call/call_93-5_ch2.htm.

[6] National System for Geospatial Intelligence, *Geospatial Intelligence (GEOINT) Basic Doctrine,* Publication 1–0, National Geospatial Intelligence Agency, September 2006.

[7] Attributed to Dwight D. Eisenhower as quoted by Richard Nixon in *"Khrushchev," Six Crises,* Doubleday, 1962.

[8] Clausewitz, C., *On War,* Book 2, Princeton, NJ: Princeton University Press, 1989, Chapter 2, Paragraph 24.

[9] Medby, J. J., and R.W. Glenn, 2002, *Street Smart: Intelligence Preparation of the Battlefield for Urban Operations*, Santa Monica, CA: RAND Corporation, 2002.

[10] Field Manual 3–24, *Counterinsurgency*, Headquarters, Department of the Army, 2006, Forward, http://www.usgcoin.org/ligrary/doctrine/COIN-FM3-24.pdf.

[11] The Evolution and Importance of Army/Marine Corps Field Manual 3–24, *Counterinsurgency,* Small Wars Journal, http://smallwarsjournal.com, June 27, 2007.

[12] Vick, A., *Aerospace Operations Against Elusive Ground Targets,* Santa Monica, CA: RAND Publications, 2001, pp. 45–46.

[13] Birkel, P. A., "Terrain Trafficability in Modeling and Simulation," MITRE, Technical Paper Sedris- 2003–1, MITRE Corporation, McLean, VA.

[14] Donlon, J. J., and K. D. Forbus, "Using a Geographic Information System for Qualitative Spatial Reasoning About Trafficability," *Proceedings of QR99*, Scotland, June 1999.

[15] Antony, R., *Principles of Data Fusion Automation*, Norwood, MA: Artech House, 1995.

[16] Joint Publication 2–01.3 *Intelligence Preparation of the Operational Environment*, Office of the Chairman of the Joint Chiefs of Staff, June 16, 2009.

[17] http://www.zeltech.com.

[18] Joint Publication 1-02, *Department of Defense Dictionary of Military and Associated Terms,* Office of the Chairman of the Joint Chiefs of Staff, 2001, as amended through June 13, 2007, p. 265.

[19] Hall, D. L., and R. Linn, "Comments on the Use of Templating Techniques for Multisensor Data Fusion," *Proceedings of the Third Tri-Service Data Fusion Symposium*, Applied Physics Laboratory, Johns Hopkins University, Laurel, MD, Vol. 1, May 1989, pp. 345–354.

[20] Joint Publication 2–0, *Joint Intelligence*, Office of the Chairman of the Joint Chiefs of Staff, June 16, 2007.

[21] Secretary of the Army and Chairman of the Joint Chiefs of Staff, *Serving a Nation at War* Washington, D.C.: U.S. Government Printing Office, no date, 2.

[22] http://www.magellangps.com/.

[23] http://www.accuweather.com/.

[24] http://www.otherweather.com/.

[25] http://www.ngdc.noaa.gov/mgg/topo/topo.html.

[26] http://www.ngdc.noaa.gov/mgg/topt/globeget.html.

[27] http://gliff.org/usgs/.

[28] http://www.tectonique.net/tectask.

[29] http://nhd.usgs.gov.

[30] *Hydrology Handbook*, 2nd ed., American Society of Civil Engineers, 1996.

[31] http://www.teleatlas.com/index.htm.

[32] http://www.navteq.com/.

[33] http://www.tomtom.com/index.php?Lid=4.

[34] http://www.grid.unep.ch/data/links_envt_data_sources.php.

[35] http://www.epa.gov/epahome/data.html.

[36] www.vesseltracker.com.

[37] http://www.world-nuclear.org/info/inf01.html.

[38] http://geoheat.oit.edu/directuse/power.htm.

[39] http://www.gsmworld.com/roaming/gsminfo/index.shtml.

[40] http://www.antennasearch.com/.

[41] http://www.sciencedaily.com/releases/2009/07/090706161304.htm.

[42] http://www.banksdaily.com/world/finance.

[43] http://www.gallup.com/home.aspx.

[44] https://www.cia.gov/library/publications/the-world-factbook/.

[45] http://www.state.gov/.

[46] http://data.un.org/.

Chapter 9

Information Fusion for Civilians: The Prospects of Mega-Collaboration

Mark S. Pfaff, Christine M. Newlon, Himalaya Patel, and Karl F. MacDorman, School of Informatics, Indiana University, Indianapolis

Current research in information fusion is redefining the role of human participants. This human-centered approach has highlighted the public's potential to observe and report information and to analyze complex problems. This is especially true for problems embedded in social activities and social networks.

9.1 INTRODUCTION

In this chapter we explore a bottom-up perspective on information fusion in the civilian layer by reviewing how ad hoc networks of volunteers have formed and functioned to address large-scale problems and by discussing how information and communications technology (ICT) could be designed to support this activity, which we refer to as *mega-collaboration*. The goals are not only to better facilitate civilian responses to crises, but also to interface these bottom-up networks with the top-down structures of military and governmental agencies.

9.1.1 Informational Crowdsourcing

The concept of crowdsourcing explores the analytic and information gathering power of individuals and groups outside of formalized military or governmental structures [1]. Although information fusion recognizes the potential value of the civilian layer (also called the H-Space [2]), it typically casts civilians as resources in a top-down structure. From this top-down perspective, citizens may be passively observed, actively solicited for information, or called openly to investigate a problem.

A different kind of organizational structure to consider is that of teams that form from the bottom up through the efforts of civilian volunteers spontaneously responding to a problem or event. These ad hoc networks can form to gather information, share knowledge, and take action independently of the command and control structures of official response agencies. There are growing numbers of examples of this phenomenon to study, such as the civilian responses to the September 11, 2001, attacks on the World Trade Center buildings in New York City [3], the 2004 Indian Ocean tsunami, Hurricane Katrina in 2005 (both in [4]), and the April 2007 shootings at Virginia Tech [5]. In these and other cases, the public demonstrated remarkable creativity and agility in gathering, processing, and disseminating information by whatever means were available—from Internet posts to paper flyers [4]. This rapid summoning of energy enabled these nascent groups to take effective action on problems even before official responders had arrived on the scene. Civilians on the ground may be more likely to know what and where the problems are and the location and means of obtaining needed resources. However, as these ad hoc networks are formed rapidly with opportunistic appropriation of communications technology, they become resistant to hierarchical organization and structured communication with official agencies. This resistance is particularly pronounced when trying to bridge official agencies (e.g., FEMA or the National Guard) with civilian efforts.

9.1.2 Mega-Collaboration

The term *mega-collaboration* appears to have been coined by Nielsen [6] to describe activity on the Web in which independent actions from millions of people (a "city of strangers") acting in their own interest collectively create a productive environment. This is quite unlike typical collaboration in which team members know each other and share explicit objectives. We prefer to add this higher level of goal orientation and self-organization to Nielsen's concept, particularly in light of evidence that strangers facing a common problem can and will exploit Internet-based technologies (e.g., social networking sites, blogs, and chat rooms) to seek potential associates, form groups, share information, negotiate strategies, and take action. In practice, mega-collaboration converges toward more typical collaboration, though still on a vast, potentially global, scale.

This phenomenon is highly visible in the crisis-response domain, which is the focus of the present discussion. However, mega-collaboration can also be applied to other situations where self-organized public activity should partner with official administrative structures, such as between neighborhood crime watch organizations and municipal police forces. Furthermore, mega-collaboration is not necessarily constrained by local or temporary disabling of communication media. First, although communications infrastructures can be heavily compromised by some crises, especially natural disasters, the aftermath can last months or years, far surpassing the recovery time of the communications networks. Local telecommunications networks are likely to be restored long before the recovery is complete, while people are still assessing the damage and casualties, locating resources, and reestablishing acceptable living and working conditions. Meanwhile, the rest of the global community continues to gather information and organize resources. Mega-collaboration extends long after its precipitating event and across a far wider area.

Mega-collaboration shares many research threads with traditional data fusion, including situation assessment, group cognition, and common operational pictures (COP) as applied toward complex problem solving and resource allocation. Research on these mega-collaborative processes aims to understand these problems more fully and to explore potential sociotechnical solutions. Drawing from the living laboratory approach [7], we have identified three main thrusts in this area:

1. Understanding the social processes of technologically mediated communication of ad hoc teams in response to complex, large-scale events. A better understanding of these social processes, especially the disjunctions between them, is essential to inform the development of flexible and transparent systems that afford improved situation awareness and facilitate rapid and effective team cognition. This can be achieved by conducting field work, examining case studies, running experiments with volunteers in simulated task environments, and linking theoretical approaches from social psychology, industrial-organizational psychology, and human–computer interaction.

2. Identifying procedural and technological interventions to address the gaps identified above. Armed with knowledge about the individual and team activities to be supported, existing technology may be reevaluated, and new technosocially appropriate systems can be proposed and developed. Successful results will come from an approach that is simultaneously user-centered and group-centered.

3. Testing prototypes of tools to support mega-collaboration with human volunteers. Armed with an understanding of social processes involved in large-scale disaster response, impediments to their smooth functioning, and promising procedural and technological interventions, innovative tools can be developed, tested, and deployed to facilitate communication among

potential volunteers, team formation, and collective action and to enable better integration of the civilian layer with disaster response organizations.

9.2 THE ROLE OF ICT IN DISASTER RECOVERY

The advance of information and communication technology (ICT) has added a new dimension to research on disaster relief in terms of both potential problems and potential solutions. Concurrently, the evolving discipline of informatics has been leading to a more rigorous consideration of the implications of ICT development for collaborative information gathering and other activities.

9.2.1 New Dimensions in Research in Disaster Recovery

The knowledge and resources needed to confront a crisis are often distributed, politically and physically, among multiple agencies and geographic locations. This situation has led some crisis-response researchers to call for a distributed decision-making network for the management of mega-disasters [8]. However, current technological support for mega-scale distributed collaboration is inadequate [9]. Responders need better support through more effective interfaces to help them convert masses of distributed data into appropriate action.

Hurricane Katrina and other recent mega-disasters have spurred a new kind of mega-collaboration in which thousands of people respond to a crisis by spontaneously working together via the Internet [10]. Ordinary citizens and their grassroots organizations have rapidly connected volunteers, donors, and aid recipients by updating blogs, electronic mailing lists, and bulletin boards. These technologically empowered volunteers should be managed as part of the overall response to a disaster to avoid adding to the chaos. However, because they are geographically dispersed and demographically diverse [3], they present a serious management problem.

A trade-off exists between the benefits of command-and-control structures efficiently delivering services under extreme conditions and of thousands of spontaneous volunteers and emergency organizations responding creatively to unforeseen problems [11]. Grassroots self-organization among the affected population contributes to the adaptability, creativity, and improvisation that are critical to the success of the relief effort [12].

This line of thought has developed into a call to action in a paper on *collaborative adhocracies* by Mendonça, Jefferson, and Harrald [13]. This call specifically targets ICT designs that rely on outdated approaches to disaster response. They instead propose *emergent interoperability* as a more appropriate approach to the design of ICT for disaster response. By this they mean a structured methodology for making use of a wide range of available ICTs selected in real time to support both individuals and groups involved in the emergency response.

In a similar vein, Denning [3] described how multiorganizational networks form after a disaster and the factors that determine their success. A hastily formed network (HFN) is a rapidly established network of people from different communities who work together to achieve an urgent mission in a shared conversation space. The HFN encompasses both the communication system and how users interact within it. Creating well-functioning HFNs poses a challenge for ICT design. After examining the responses to both the 2001 World Trade Center attacks and Hurricane Katrina, Denning observed, "[The] effectiveness of the HFN rests on the quality of the conversation space established at the outset" [3, p. 17]. If participants can agree on interaction rules and reach a consensus on the definition of the problem, the likelihood of success greatly increases. This process of negotiation is what mega-collaboration tools should be designed to support.

9.2.2 Calls for ICT Innovation for Disaster Collaboration Support

Palen et al. [5] documented the public's use of social networking during the Virginia Tech shooting in April 2007. Private citizens (many of whom were located far from Virginia) performed much unsolicited work in compiling a list of victims and connecting students, staff, and faculty with distant worried relatives. In fact, an HFN had already compiled a complete list of victims before the officials in charge at the scene had released theirs. This is yet another example of ICT-enabled collaboration and information gathering by the public outstripping the official response (also see [14, 15]). Although the public's newfound agility for self-organization might be seen as beneficial, it also invites potential dangers if the gap between public and bureaucratic agility continues to widen. Palen, Hiltz, and Liu [16] addressed this issue by describing ethnographic studies on the World Trade Center attacks, the London Tube bombings, Hurricane Katrina, the California wildfires, the SARS epidemic, and various earthquakes around the world. These studies show that the public usually responds first to a crisis and does not relinquish its role once the official effort begins. This has been true even when the only available methods of response and communication were digging with bare hands and posting paper flyers. Therefore, it is not surprising that the public has led the way in adopting novel technology applications in times of crisis. Widely available ICT advances challenge the conventional models used by government planners and will require a new relationship between official responders, nongovernmental organizations (NGOs), and the public [17, 18]. These advances enable new designs for software tools that foster effective collaboration between official responders and private citizens.

9.3 INFORMATION FUSION THROUGH MEGA-COLLABORATION PROCESSES AND TOOLS

Information fusion at the scale of large civilian populations introduces multiple challenges of scale, correlation, normalization, and resource management at the levels of both human and technical capabilities. New computational approaches are being explored for many facets of mega-collaboration.

9.3.1 Social and Cultural Processes

In general, collaboration demands that individual participants function as a team, traversing the team-building stages of *forming, storming, norming,* and *performing* [19]. To succeed at team-building, teammates must combine their individual mental models of the problem into a team model. This involves both the convergent processes of information pooling and cognitive consensus and the divergent processes of specialization and transactive memory (i.e., transmission of the cooperative information to the appropriate expert [20]). Therefore, large-scale collaboration in a distributed environment requires an interface that captures individual mental models and facilitates the negotiation of team models. The goal of mega-collaborative systems is to aid in the comparison and merging of these models such that a hierarchy of consensus, organized tasking, and a common operational picture emerge from this expansive community of individual participants.

The design of a large-scale collaborative interface poses social, psychological, and technological research questions. The formation of mental models is a dynamic process involving both the individual and the situation. Capturing such models requires a flexible interface capable of representing many different kinds of entities and relations. An even greater challenge is facilitating the model-negotiation process among a dispersed and heterogeneous team. These challenges are particularly daunting, because they must be met for a team of thousands. Ongoing research offers potential solutions [21–23]. This chapter provides a synthesis of these approaches for developing a tool for managing mega-disasters.

Research on team dynamics has increased our understanding of cooperation, suggesting new tools for online collaboration. Ess and Sudweeks [24] and Hewling [25] described how virtual-group participants from different organizational cultures negotiate a new *third culture*. This new culture is created out of the participants' unique online encounters. Certain individual personality traits have been identified that affect interpersonal interactions, such as conscientiousness, agreeableness, and neuroticism [26]. Several studies have been conducted on virtual teams [27, 28] and extreme teams with several hundred members. The latter are typically seen in emergency response situations [29, 30].

9.3.2 Collaboration Management

To organize human-reported information into a meaningful conversation, some level of collaborative administration is necessary. Information management challenges must also be overcome to link the civilian layer with tactical operations. One approach to address this is *collaboration engineering*, which facilitates the decomposition and design of repeatable collaboration processes for teams working on high-value collaborative tasks [31]. The goal is to provide neutral guidance and structure to the collaborative process without requiring a trained meeting facilitator. Collaboration engineering supports team modeling by constructing a negotiation process from a sequence of individual process segments called *thinkLets* [32]. A thinkLet is "a named, packaged facilitation technique captured as a pattern that collaboration engineers can incorporate into process designs" [32, p. 1]. Collaboration processes divide into several goal categories: divergence, reduction, clarification, organization, evaluation, and consensus building. By breaking up the team activity into segments, each with one of these goals, it is possible to build a negotiation process that captures all the ideas contributed while allowing participants to focus quickly on what is important.

In field trials novice group leaders found it relatively easy to master and execute thinkLet-based process designs. Novices led these processes without the weeks or months of apprenticeship typically needed to learn collaboration facilitation [33, 34]. Collaboration engineering researchers have employed the thinkLet pattern language to design a number of collaboration processes that have been implemented successfully in commercial, government, and military organizations for such applications as crisis response training and operational execution [35], biocontainment [36], and policy analysis [37].

Collaboration engineering has thus far focused on generating text-based dialogues. The next step to support mega-collaboration is to extend thinkLets to complex mental models stored in a relational database. The application of collaboration engineering to distributed environments is just starting. An exploratory study using Groove as a distributed collaboration platform illustrated the potential of thinkLets to support distributed teams in the effective execution of a requirements definition task [38]. This study also showed a variety of important areas of future research, such as the degree to which thinkLet-based processes must be adapted to ICT, the design and evaluation of effective thinkLets for distributed collaboration, and the nature of leadership in temporary distributed teams.

9.3.3 The Contribution of Artificial Intelligence

Even with these approaches, managing the development of team models on the massive scale of a mega-disaster will require artificial intelligence. Several studies have documented the success of large-scale team management using autonomous software agents. In each case, the team was divided into subteams and managed by communications among the agents via a small-worlds network [39–41]. This kind of process can manage the comparison and synchronization of models by subteams of users, thereby facilitating information pooling, cognitive consensus, and transactive memory. A mixed-initiative interface augmented by data-mining techniques would allow both humans and software agents to extract actionable information from the project database.

Collaboration among autonomous software agents, and between these agents and humans, has shown great potential for disaster response [42]. However, the research mentioned earlier on small-world networks involved simulations in which autonomous software agents adopted theoretical roles representing human actors. Instead of replacing human actors with sense-making software agents, it is possible to employ human teams for sense-making in an agent-managed network. This structure combines the strengths of humans for observation and inference and the high availability of computers for rapid comparison and organization of information. Such an alliance would allow the agents to monitor the need for collaborative action and to broker both the information exchange and the collaborative sequence in a manner aligned with the thinkLet designs described above.

9.3.4 Individual and Team Interfaces

A user-friendly interface and an intuitive functionality are essential to allow individuals to connect over the Internet, discuss important issues, and develop teams to take action. As teams form, the interface should support the development of individual and team mental models via the front-end input and output, as well as the back-end team management mechanisms. This is necessary to organize the goals and actions that are of common interest to the participants. This interface should enable teams to organize a robust picture of their shared data while automatically creating the data structure to manage it (i.e., the interface maintains a shared meaning in the data without forcing users to add semantic markup). Exploring this common picture together as a team-building exercise encourages a shift from competitive to cooperative behavior [27].

However, there are also several constraints to consider, such as gaining access to the tool, developing sufficient interest to use the system and participate with other teams, and understanding both the subject matter and the system interface. All of this must be performed under conditions that may be highly stressful. Although individuals have employed existing Web-based tools (such as Facebook, MySpace, Second Life, Flickr, and others) for ad hoc information fusion in past crises, these systems, designed for social tasks that are not mission-critical, have proven unwieldy and inefficient for crisis response [5]. This demonstrates an interest in participation and a need for new online venues and meeting places designed to support grassroots information fusion. We propose that this can be achieved through a system that allows individuals to share mental models of the situation and provides support to visualize, compare, and merge these models for organized collaborative team efforts.

Preliminary tests of our current prototype interface indicate that sophisticated interface design can enable a tool to guide individuals through the definition of their mental models [10, 23, 43]. As frameworks for network application development have matured, capturing the users' concepts and routing them to a back-end database have become easier through a process mediated by middle-tier business logic. These concepts are restructured into a set of entities and relations that can be categorized as events, goals, tasks, roles, actors, and resources [44]. In addition, the online conversation surrounding this process can be captured and preserved in its context [45]. The interface must support users in converting their thoughts into representations that can be compared with those of their teammates.

9.4 IMPLICATIONS FOR DESIGN AND DEVELOPMENT

While it is obvious that the issues of scale are central to the mega-collaboration problem, the relevant tools must also be designed for use in extreme circumstances such as earthquakes or hurricanes. The state of the field is largely driven by assessments of who will be using the tool and which capabilities will be required given the wide variation in users' backgrounds, assumptions, and training.

9.4.1 Current Experimental Work

Our current experimental work is inspired by the recognized need for coordination among spontaneous grassroots responders. A long-term goal is a deployable Internet-based mega-collaboration tool (MCT). The central concept behind the tool is that a massive problem (e.g., rebuilding a demolished home) can be incrementally engaged by multiple small subteams ("we need to find more lumber"), each developing a model to define part of the problem through a protocol consisting of collaboration engineering thinkLets. Consolidating these models in agent-augmented compare-merge playoff sessions will allow mega-teams to agree on the definition of the problem and coordinate effective action. This enables a paradigm shift in employing Web 2.0 technologies to increase the effectiveness of crisis response, allowing for larger teams and a wider range of topics. It is the mental model refinement via compare-merge sessions, the scalability of the mega-teams, and the computational swiftness by which the agents facilitate these collaborative actions that demonstrate how categorically different the approach is from those of traditional groupware applications.

As an example of this communication process, representatives from different subteams would use this tool to resolve conflicts by negotiating with one another. For instance, if two teams plan to evacuate the same church, each will send representatives to a negotiating team, bringing with them a data structure identifying the church, the goal of evacuating it, and other information that their group has gathered about the situation. For consolidation purposes, the details from each model will be combined, and any duplicate items will be eliminated. The team representatives can then negotiate via the chat room what resources are still needed.

9.4.2 The Mega-Collaboration Tool

The current work builds on specifications that have been developed over a series of preliminary studies [22, 23, 43]. A prototype tool has been constructed and tested using light-weight, browser-based, open-source software. Tests of the tool have determined that it enhances an online team's effectiveness as measured by how well it defines its problem space and comes to agreement on what actions to take. However, these tests have only examined within-team behaviors and attitudes. Future work will evaluate the tool's performance when multiple individual teams are combined into a mega-team.

Cognitive walkthroughs in the tool's preliminary design stages indicate that the problem-definition task impedes use. To overcome this, a problem-definition protocol was introduced that enables each teammate to form an individual mental model of the problem and then to negotiate a team model. As the teammates work, the tool reflects their progress by adding structures to the database, which it draws on to create visualizations for the team. A chat window lets teammates communicate during any of the coordination stages.

The database supporting this activity is sufficiently general that teammates can flexibly create their own problem definitions [10, 43]. Although the data-definition protocol encourages teammates to define their problem in terms of *events*, *goals*, *tasks*, *resources*, and *roles*, the database treats each of these definitions as a generic entity. The name and description of each entity are therefore added to the entity table. Because one person's goal may be another person's event or role, a situation table identifies the particular situation in which a given entity is being represented. This allows entities to be combined if they are found to be identical, without losing the situational differences between the two definitions. The database also has a relations table that allows for the relationships among the different entities to be expressed. The result has been a reconfigurable database that can store very complex data.

9.4.3 Use Cases

To further define the MCT concept, we developed a number of theoretical user profiles and use cases drawn from users and events documented following Hurricane Katrina. The representative users for which we developed profiles are summarized in Table 9.1.

These demonstrate the diverse needs resulting from a major disaster, which point toward effective strategies for how the technology could meet those needs. It was immediately apparent that individuals and groups would require customized or customizable interfaces. However, all the information should be drawn from a common database. Further, the automated agents would have to act independently to coordinate the asynchronous information gathering and model development processes among the groups.

Table 9.1
User Profiles

Type	*User*	*Motivating Goal for Use*
Local Emergency Responders	District Fire Superintendent	Determination of Priorities
Volunteer Labor Organizations	Firefighters' Union Coordinator	Resource Coordination
Nonprofit Aid Organizations	Red Cross Coordinator	Resource Coordination
Military Organizations	National Guard Coordinator	Response Activity Tracking
Federal Emergency Responders	FEMA Coordinator	Jurisdiction Coordination
Concerned Common Citizens	Store Manager	Resource Donation
Volunteer Workers	Social Worker	Resource Donation
Volunteer Experts	Computer Expert	Technology Donation
Affected Individuals	Relative	Rescue of Family Members

9.4.4 Required Features

Typical online collaborative actions would have to be supported, such as basic security features and account management, as well as a number of different possible interactions between users. These are presented in Table 9.2.

This initial set of user profiles and interaction requirements was developed into a set of preliminary specifications and a concept prototype [10]. A more detailed paper prototype was refined during a series of focus group sessions. Their results led to the first working prototype of the MCT, which was subsequently used to refine the team-building interface and test the effect that negotiation of mental models had on the team decision-making process [22]. The initial version of the MCT was developed using an AJAX-based interface with a PHP and MySQL back end. An open-source database structure was selected to maximize the future connectivity of the MCT with other information fusion systems.

Early findings strongly supported the theoretical underpinnings of information fusion at the grassroots level. Conversation analysis of the tests indicated that teams with an emergent bottom-up development of leadership produced more successful action plans. Teams also preferred to have a single leader instead of sharing leadership among all teammates. Individually developed models were generally disorganized lists of information and ideas, but the subsequent compare-and-merge process proved highly effective in resolving all of that information into a complex hierarchical group model. This complex information structure was maintained when the model was drawn into an action plan. However, the action plans from the control teams, which had no access to the modeling functionality, remained as disorganized lists of ideas.

A subsequent two-part study examined participants' experiences using the first and second generation modeling interfaces. In the first part of the study, we

Table 9.2

Interaction Requirements

ID	*Interaction*	*ID*	*Interaction*
1	Find Site	10	Develop Mental Models
2	Use Site	11	Negotiate Group Models
3	Find Area of Interest	12	Vote
4	Participate	13	Take Turns
5	Converse	14	Exchange Information and Resources
6	Create Team	15	Form Teams of Agents
7	Join Team	16	Agent-Mediated Playoffs
8	Leave Team	17	Intergroup Negotiation
9	Disband Team	18	Provide Help

gathered ideas for revising the first generation interface. Participants were assigned randomly to an *interface team* or a *control team*. Interface teams completed a brief tutorial and began the model-building process for an assigned problem (namely, creating a business plan). Control teams worked on the same problem space using the interface's text-chat functionality, which was the only component of the interface available to them. Both teams experienced difficulties adapting their problem space to the interface. Following the initial tests, we introduced a new front-end, built-in Adobe Flex (Figure 9.1, [23]).

Because civilians new to mega-collaboration may know little about the constraints and workflow of the software, the second part of the study used 10 participants who had no prior experience with either interface. We tested participants individually, assigning each person to one of the two interfaces and giving each a list of commonly performed tasks. After completing the tasks, participants evaluated the interface along a diverse set of usability factors, including information quality, interface quality, interface learnability, interface aesthetics, and emotions elicited by the interface. Participants also responded to items regarding team-creation functionality, input and output interfaces, and the model-building process [23].

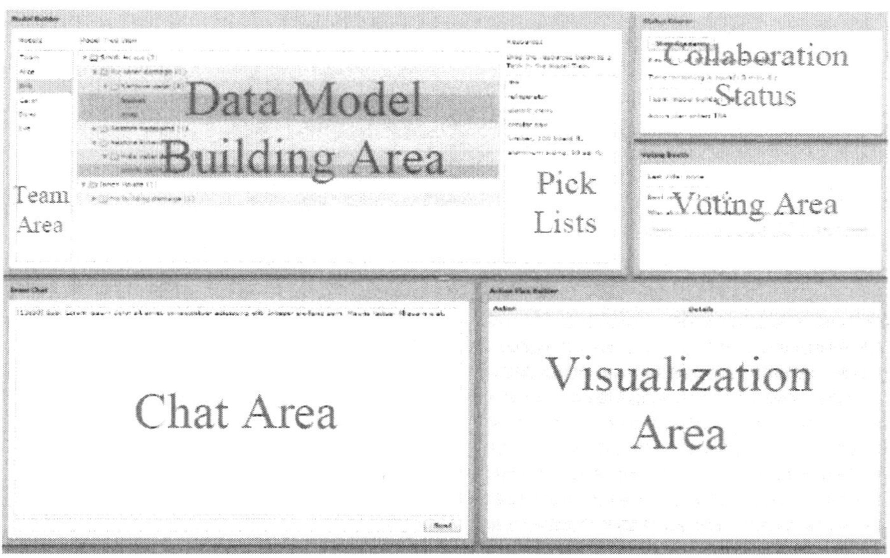

Figure 9.1 Second generation of the mega-collaboration prototype interface.

The usability studies indicated that the interface's flexibility is vital to the successful practice of mega-collaboration. The most common usability-related discrepancies between the two prototypes involved data input, visualization, and data categorization. Participants believed that the second generation interface was significantly better suited for first-time users. They also believed the second generation interface was more enjoyable and more tasteful. The second interface was associated more strongly with the descriptive term *energetic* and the emotional term *frenzied* (versus *sluggish*). Participants reported that the forced categorization of each mental-model object as an event, goal, task, role, or resource was too rigid. They requested more ways to manipulate their data, including cut-and-paste, importing from external data sources, and temporal organization. In poststudy interviews, participants requested the ability to work with partial data hierarchies by attaching, detaching, and reorganizing them. The strengths of a shared predefined structure for mental-model objects should be considered in relation to the impositions that the structure makes on its users. Even if everyone in a group derives his or her solution-finding process from the bottom up, individual collaborators may construct mental models as narratives or in iterative revisions. Expecting a group to work solely in one direction (e.g., from the problems to the goals or vice versa) is inefficient at best and counterproductive at worst.

Usability results help guide the development of the interface, but behavioral observations are especially valuable in revealing how individuals come together and create structures—team structures as well as information structures—in a relatively free-form and self-guided fashion. With further study, the MCT's efficacy can surpass that of repurposed social networking tools for enabling civilian information gathering in response to a crisis.

9.5 CONCLUSIONS

The mega-collaborative approach to information fusion is innovative in the following respects:

- Citizen volunteers are encouraged to develop their own problem-definition models and are supported in the negotiation and consolidation of these models in virtual teams.
- The compare/merge features will leverage the strengths of people at conceptualizing information and the strengths of computers at managing information. This synergy will be accomplished by having the participants construct and negotiate their own models and by having autonomous software agents track and route the data.

- Formats of collaboration engineering that were formerly based on unstructured text will be adapted to complex, hierarchical data structures supported by the autonomous agents.
- The proposed tool will transform the multidimensional, heterogeneous data resulting from disasters into formalized data structures, thereby allowing distributed decision-making networks to be integrated with centralized command structures.
- By allowing parallel, asynchronous data flow, the proposed tool will scale the virtual teams to sizes that previously could not be handled efficiently.

The goal of supporting mega-collaboration is too ambitious for any single research program to pursue comprehensively, but this chapter has presented examples to encourage more thinking and research in this complex area.

We believe that these technologies are best tested in environments that effectively mimic the real-world conditions for which the tools are designed. Current user tests have so far been conducted with static scenarios, but as the MCT becomes more stable and powerful, tests will be conducted using NeoCITIES [46], a computer-based scaled world simulating the situation assessment and resource allocation tasks of distributed emergency crisis management teams. In NeoCITIES, the group activity consists of distributed individuals jointly gathering information about emergency events, allocating resources to address these events, and detecting emerging threats and patterns of activity from an underlying scenario. This experimental approach provides a holistic assessment of distributed cognition with real-time performance, tool use, and team communication measures.

Information fusion of human-reported data presents a host of computational difficulties, such as the descriptive subjectivity of reports (compared to the calibrated accuracy of physical sensors), the expression of the information in a natural language, and the general autonomy of the actors in the system. Thus, machine readability and manageability of the ad hoc team network activities have become priorities. As mentioned previously, we are particularly interested in the potential impact of augmenting these processes through artificially intelligent mixed-initiative agents to enhance situation awareness and process management. A reliable instrument for converting human-supplied data into easily accessible information will improve the impact of the decisions made by the autonomous agents and the effectiveness of the overall response to the crisis.

The expected outcome of this work is that responders to a crisis will be able to locate information in their area of interest or expertise and contribute additional information, resources, and decision-making power to address the crisis. The results are expected to enhance substantially the effectiveness of disaster response as well as provide valuable insight into the processes by which ad hoc teams become mega-collaborative organizations.

The successful development of tools for mega-collaboration will enhance society's ability to respond not only to disasters, but also to any problem that requires broad understanding and agreement. The principles discussed in this chapter can be applied to almost any team-based project and may inspire new methods of decentralized decision-making and coordination.

As a final comment, it should be noted that this chapter has not discussed issues related to the privacy policies and issues related to the new capabilities of mega-collaboration. The American Civil Liberty Union (ACLU) Web site (www.aclu.org), for example, has a number of reports regarding the growth of surveillance technology and the lag in associated national policies and procedures to protect citizen privacy. In addition, a recent article by Shilton [47] provides an introduction and discussion of issues related to how participatory sensing and collaboration provide challenges to individual and collective privacy. While it is beyond the scope of this chapter (or book) to address these issues, there must certainly be extensive discussions and investigations regarding the balance between technology which enables public surveillance and crowd-based collaboration for reporting news, events, and activities and individuals' rights to privacy.

References

[1] Howe, J., *Crowdsourcing: Why the Power of the Crowd Is Driving the Future of Business*, New York: Crown Publishing Group, 2008.

[2] Hall, D. L., et al., "A Framework for Dynamic Hard/Soft Fusion," *Proceedings of the 11th International Conference on Information Fusion*, Cologne, Germany, 2008, pp. 1–8.

[3] Denning, P. J., "Hastily Formed Networks," *Communications of the ACM*, Vol. 49, No. 4, 2006, pp. 15–20.

[4] Palen, L., and S. Liu, "Citizen Communications in Crisis: Anticipating a Future of ICT-Supported Public Participation," *Proceedings of the SIGCHI Conference on Human Factors in Computing Systems*, San Jose, CA, 2007, pp. 727–736.

[5] Palen, L., et al., "Crisis Informatics: Studying Crisis in a Networked World," *3rd International Conference on e-Social Science*, 2007.

[6] Nielsen, J., "Community Is Dead; Long Live Mega-Collaboration," 1997, from www.useit.com/alertbox/9708b.html.

[7] McNeese, M., et al., "Encountering Computer-Supported Cooperative Work Via the Living Lab: Application to Emergency Crisis Management," *Proceedings of the 11th International Conference on Human-Computer Interaction,* Las Vegas, NV, 2005, pp. 1–8.

[8] Harrald, J., and T. Jefferson, "Shared Situational Awareness in Emergency Management Mitigation and Response," *Proceedings of the 40th Annual Hawaii International Conference on System Sciences* (HICSS'07), 2007, p. 23b.

[9] Denning, P. J., and P. Yaholkovsky, "Getting to 'We,'" *Communications of the ACM*, Vol. 51, No. 4, 2008, pp. 19–24.

[10] Newlon, C., and A. Faiola, "Support for Mega-Team Collaboration with Cultural Tools: Providing a Framework for the Dynamic Development of Team Culture," *Proceedings of the 12th International Conference on Cultural Attitudes Towards Technology and Communication*, Tartu, Estonia, 2006, pp. 235–254.

[11] Harrald, J. R., "Agility and Discipline: Critical Success Factors for Disaster Response," *The ANNALS of the American Academy of Political and Social Science*, Vol. 604, No. 1, 2006, pp. 256–272.

[12] Dynes, R. R., "Community Emergency Planning: False Assumptions and Inappropriate Analogies," *International Journal of Mass Emergencies and Disasters,* Vol. 12, 1994, pp. 141–158.

[13] Mendonça, D., T. Jefferson, and J. Harrald, "Collaborative Adhocracies and Mix-and-Match Technologies in Emergency Management," *Communications of the ACM*, Vol. 50, No. 3, 2007, pp. 44–49.

[14] James, A. M., and T. Rashed, "In Their Own Words: Utilizing Weblogs in Quick Response Research Learning from Catastrophe Quick Response Research in the Wake of Hurricane Katrina," Boulder, CO: University of Colorado-Boulder, Natural Hazards Center, 2006, pp. 71–96.

[15] Schneider, S. M., and K. A. Foot, "Crisis Communication and New Media: The Web After September 11," in P. N. Howard and S. Jones, (eds.), *Society Online: the Internet in Context,* Thousand Oaks, CA: Sage Publications, 2004, pp. 137–154.

[16] Palen, L., S. R. Hiltz, and S. B. Liu, "Online Forums Supporting Grassroots Participation in Emergency Preparedness and Response," *Communications of the ACM*, Vol. 50, No. 3, 2007, pp. 54–58.

[17] Currion, P., C. de Silva, and B. Van de Walle, "Open Source Software for Disaster Management," *Communications of the ACM*, Vol. 50, No. 3, 2007, pp. 61–65.

[18] McNeese, M., et al., "Multiple Vantage Points of the Common Operational Picture: Supporting Complex Teamwork," *Proceedings of the 50th Annual Meeting of the Human Factors and Ergonomics Society*, 2006, pp. 467–471.

[19] Tuckman, B. W., "Developmental Sequence in Small Groups," *Psychology Bulletin*, 1965, pp. 63384-63399.

[20] Mohammed, S., and B. C. Dumville, "Team Mental Models in a Team Knowledge Framework: Expanding Theory and Measurement Across Disciplinary Boundaries," *Journal of Organizational Behavior*, Vol. 22, No. 2, 2001, pp. 89–106.

[21] Klein, G. L., M. S. Pfaff, and J. L. Drury, "Supporting a Robust Decision Space," *Proceedings of the AAAI Spring Symposium on Technosocial Predictive Analytics*, Menlo Park, CA, 2009.

[22] Newlon, C., K. MacDorman, and P. Scerri, "A New Model for Mega-Collaboration," *First International Workshop on HCI for Emergencies*, 2008.

[23] Newlon, C., et al., "Mega-Collaboration: The Inspiration and Development of an Interface for Large-Scale Disaster Response," *Proceedings of the 6th International Information Systems for Crisis Response and Management* (ISCRAM 2009), Göteborg, Sweden, 2009.

[24] Ess, C., and F. Sudweeks, "Culture and Computer-Mediated Communication: Toward New Understandings," *Journal of Computer-Mediated Communication*, Vol. 11, No. 1, 2005, pp. 179–191.

[25] Hewling, A., "Culture in the Online Class: Using Message Analysis to Look Beyond Nationality-Based Frames of Reference," *Journal of Computer-Mediated Communication*, Vol. 11, No. 1, 2005, article 16.

[26] Mount, M., M. Barrick, and G. Stewart, "Five-Factor Model of Personality and Performance in Jobs Involving Interpersonal Interactions," *Human Performance*, Vol. 11, No. 2 and 3, 1998, pp. 145–165.

[27] Farnham, S., et al., "Structured Online Interactions: Improving the Decision-Making of Small Discussion Groups," *Proceedings of the 2000 ACM Conference on Computer Supported Cooperative Work*, Philadelphia, PA, 2000, pp. 299–308.

[28] Powell, A., G. Piccoli, and B. Ives, "Virtual Teams: A Review of Current Literature and Directions for Future Research," *SIGMIS Database*, Vol. 35, No. 1, 2004, pp. 6–36.

[29] Scerri, P., et al., "Allocating Tasks in Extreme Teams," *Proceedings of the Fourth International Joint Conference on Autonomous Agents and Multiagent Systems*, 2005, pp. 727–734.

[30] Scerri, P., et al., "Scaling Teamwork to Very Large Teams," *Proceedings of the Third International Joint Conference on Autonomous Agents and Multiagent Systems*, Volume 2, 2004, pp. 888–895.

[31] Briggs, R. O., G. J. de Vreede, and J. F. J. Nunamaker, "Collaboration Engineering with ThinkLets to Pursue Sustained Success with Group Support Systems," *Journal of Management Information Systems*, Vol. 1, No. 4, 2003, pp. 31–63.

[32] de Vreede, G. J., G. L. Kolfschoten, and R. O. Briggs, "ThinkLets: A Collaboration Engineering Pattern Language," *International Journal of Computer Applications in Technology*, Vol. 25, No. 2/3, 2006, pp. 140–154.

[33] Agres, A., G. J. de Vreede, and R. O. Briggs, "A Tale of Two Cities: Case Studies of GSS Transition in Two Organizations," *Group Decision and Negotiation*, Vol. 14, No. 4, 2005.

[34] de Vreede, G. L., and R. O. Briggs, "Collaboration Engineering: Designing Repeatable Processes for High-Value Collaborative Tasks," *Proceedings of the Hawaii International Conference on System Science,* Los Altos, CA, 2005.

[35] Appelman, J. H., and J. van Driel, "Crisis-Response in the Port of Rotterdam: Can We Do Without a Facilitator in Distributed Settings?" *Proceedings of the Hawaii International Conference on System Science*, 2005.

[36] Smith, P. W., et al., "Designing a Biocontainment Unit to Care for Patients with Serious Communicable Diseases: A Consensus Statement," *Biosecurity and Bioterrorism: Biodefense Strategy, Practice, and Science*, Vol. 4, No. 4, December 2006, pp. 351–365.

[37] Enserink, B., "Creating a Scenariologic: Design and Application of a Repeatable Methodology," *Proceedings of the IEEE 36th Hawaiian International Conference on System Sciences*, IEEE, 2003.

[38] Tarmizi, H., et al., "Collaborative Engineering in Distributed Environments," *e-Service Journal*, Vol. 6, No. 1, Fall 2007, pp. 76–97.

[39] Scerri, R., J. A. Giampapa, and K. P. Sycara, "Techniques and Directions for Building Very Large Agent Teams," *Proceedings of AAMAS '05*, 2005.

[40] Scerri, P., et al., "Scaling Teamwork to Very Large Teams," *Proceedings of the 3rd International Joint Conference on Autonomous Agents and Multi-Agent Systems*, New York, July 19–23, 2004, pp. 888–895.

[41] Schurr, N., et al., "The DEFACTO System for Human Omnipresence to Coordinate Agent Teams: The Future of Disaster Response," *Proceedings of the Fourth Joint Conference on Autonomous Agents and Multi-Agent Systems,* 2005, pp. 1229–1230.

[42] Tate, A., "The Helpful Environment. Geographically Dispersed Intelligent Agents That Collaborate," *IEEE Intelligent Systems*, Vol. 21, No. 3, 2006, pp. 57–61.

[43] Newlon, C. M., A. Faiola, and K. F. MacDorman, "Building the Mega-Collaboration Interface: Behavioral, Cultural, and Cognitive Factors in Visualization Support," *Proceedings of the 12th International Conference on Information Visualisation* (IV '08), London, U.K., 2008, pp. 509–514.

[44] van der Veer, G., and M. van Welie, "Task-Based Groupware Design: Putting Theory into Practice," *Proceedings of the Conference on Designing Interactive Systems: Processes, Practices, Methods, and Techniques*, 2000, pp. 326–337.

[45] Newlon, C., "Developing a Tool for Mega-Collaboration," *New Media Consortium, Summer Conference 2007*, 2007, http://archive.nmc.org/events/2007summerconf/materials/Mega-Collaboration.pdf.

[46] McNeese, M. D., et al., "The NeoCITIES Simulation: Understanding the Design and Experimental Methodology Used to Develop a Team Emergency Management Simulation," *Proceedings of the 49th Annual Meeting of the Human Factors and Ergonomics Society*, 2005, pp. 591–594.

[47] Shilton, K., "Four Billion Little Brothers? Privacy, Mobile Phones, and Ubiquitous Data Collection," *Communications of the ACM*, Vol. 52, No. 11, November 2009, pp. 48–53.

Chapter 10

Virtual World Technologies

Dr. Bart Pursel, The Pennsylvania State University

This chapter introduces the concept of virtual world technologies emerging from the gaming community. Examples include the popular Second Life (www.secondlife.com) and OLIVE. These tools enable multiperson distributed collaboration in a common virtual environment. The environments typically allow creation of 3-D artifacts (e.g., buildings, landscapes, meeting rooms) and provide the creation of avatars to represent the collaborating participants. These environments are promising for encouraging distributed ad hoc collaboration among diverse analysts to address complex problems. Mark Bell of Indiana University [1] has provided a definition and taxonomy of virtual world terminology and a historical perspective on their evolution. Virtual world technologies support massively multiuser online games such as MMOG, MMORPG, and Metaverse.

10.1 INTRODUCTION

We explore an emerging concept in which an ad hoc community of analysts could support dynamic analysis of evolving situations. That is, just as a community of observers may be tasked (or volunteer) to provide input data of value regarding an evolving event, crisis, emergency, or other situation, we consider the possibility of a future in which a community of analysts could collaborate to analyze evolving situations using the media of virtual worlds such as Second Life or OLIVE. In this new concept, civilian or amateur analysts may seek the solution of a complex problem by collaborating over the Web analogous to the creation and maintenance of the Wikipedia. Thus, just as the national news media have begun to use civilian reporters (see, for example, CNN's iReport [http://www.ireport.com/?cnn=yes] in which amateur viewers can submit unverified videos and other reports) as observers, fusion systems could be established to solicit analytical results.

Every day, millions of people are interacting and collaborating around the world. These people form small groups, anywhere from 5 to 40 members, to test various simulations and models. Each test provides valuable data that the group uses to analyze, evaluate, and create new strategies for the next round of the simulation. These groups are interacting in an online virtual world, where they come together nightly to collaborate and overcome new challenges by collectively analyzing data, strategizing, and executing very detailed, intricate plans of action.

This probably sounds like a military training scenario, or some sort of simulated event taking place in a control center somewhere. It is not. It is what takes place every day in *World of Warcraft* (*WoW*), an online game involving over 12 million players worldwide. Teams of 40 players self-organize to take on the most challenging elements of the game, including mighty dragons and mystical gods of the elements. Some members of a 40-person team are connecting from Russia, others from Australia, and many from North America and China. Each time the group attempts to overtake one of these computer-controlled beasts, the group makes progress, learning more and more about the abilities of the beast, strategizing in real time using VoIP communications on how best to counter these abilities, and adjusting their plan of attack accordingly for the next attempt.

Imagine if this scenario could be extracted from a game like *World of Warcraft* and instead take place in a 3-D virtual world simulating emergency response scenarios, where a team of early responders with various subject matter expertise could collaborate, strategize, and execute a plan in real time to deal with simulated emergencies in realistic models of major cities. The military is already taking advantage of virtual worlds for simulated training activities using a technology platform called OLIVE, by Forterra Systems Inc. (http://www.forterrainc.com/). OLIVE and other virtual world technologies and platforms like Second Life, ProtoSphere, and Multiverse are maturing rapidly and enabling individuals to come together in rich, 3-D environments and collaborate in real time on a variety of projects.

As we move closer to the idea that everyone can be a sensor with devices like a cell phone, we are beginning to see examples of real-time data integration and mash-ups with other Web applications like Google Maps, for example. With the release of Second Life in 2003, a massive flood of virtual worlds and virtual world tools followed that are moving the genre forward, away from fantasy and game-driven worlds to business, productivity, and research applications of these spaces.

10.2 OVERVIEW OF VIRTUAL WORLDS

For many individuals, the idea of virtual worlds may seem uncommon. But virtual worlds have been in existence for quite some time. Three Web sites provide an excellent overview and background for understanding virtual world technology. The first Web site is that of Raph Koster, an industry veteran who has been part of many groundbreaking virtual world projects starting in the early 1990s [2]. The second Web site contains a presentation that Linden Labs, creators of Second Life, made available online. This presentation deals with the history of virtual worlds leading up to Second Life's launch in June 2003 [3]. The final Web site, Avatar Planet, is a Web site dedicated to "bring together the latest news about virtual worlds, both present and future, and to highlight activities and news of avatars in online virtual worlds" [4]. The site contains a high-level timeline for virtual worlds with several resources about specific virtual environments.

Virtual worlds have been in existence for many years but had not risen to the public's eye until the late 1990s. Virtual worlds can mean various things to different people. Some consider works of literature that are outside human reality a virtual world. Others consider games such as *Dungeons and Dragons* a virtual world, where the entire game is played with a pen, paper, and dice. The world itself resides in the players' imagination. For the purposes of this chapter, the focus is computer-based virtual worlds, primarily online worlds where many different players can interact synchronously. The people who inhabit these virtual worlds do so via an avatar, which is a digital representation of the player within the world that can interact with other avatars and the world's environment.

From the earliest days of networking computers together, programmers constantly toyed with gaming and virtual-world ideas via text-based games. Many two-player games were developed in the late 1960s and early 1970s, but it was not until 1974 when a game called *Empire* was released that supported 32 players. Then in 1978, Richard Bartle and Roy Trubshaw began distributing *MUD* (*Multi User Dungeon*), to friends in and around England. This was a college project for the two programmers and became wildly popular. *MUD* established the baseline mechanics for many games that are still being created and published today. By the early 1980s, many online games appeared on the scene that required the players to pay a hefty fee per hour to participate. Most games contained very primitive graphics and still relied heavily on text to drive the game forward. In the late 1980s limited 3-D games appeared, such as *A-Maze-ing*, a shooter game that ran on the Macintosh and allowed multiple players to connect synchronously. In 1986 the game *Habitat* introduced the concept of the avatar in multiplayer environments that play a major role in the game. Avatars can exist outside of virtual worlds, such as an instant messenger or a forum icon, but for the purposes of this research, the term avatar refers to a 2-D or 3-D model used in a computer game or virtual environment serving as a representation of the user [5].

LambdaMOO arrived in 1990, created by Pavel Curtis and hosted at Xerox PARC. This is a text-based virtual world where the inhabitants navigate through rooms via text input and interact with others also logged in to the world. LambdaMoo is worth mentioning for two reasons:

1. It takes the *MUD* to another level. The "MOO" in lambdaMOO stands for MUD, object oriented. What this means is that the people inhabiting the world have the ability to create interactive objects in the environment.
2. LambdaMOO is not necessarily a game, although some people have used the MOO concept to build games within the world. Because it is not a game, inhabitants use the environment for a wide variety of purposes, many being social. An extreme instance of this is when one inhabitant virtually raped another inhabitant, fueling heated debates about social issues in virtual worlds [6].

In 1996, *Meridian 59* launched. This is arguably the first massively multiplayer online (MMO) game in 3-D graphics, which pioneered the massively multiplayer online role playing game (MMORPG) genre. A year later, *Ultima Online* and *Lineage* were released. Throughout the rest of the 1990s and into the new millennium many MMORPGs were released that follow a fantasy theme involving orcs, dragons, swords, and sorcery.

Up until the mid-1990s, the player base of most online games was fairly small. Users had to pay a fixed fee per hour logged in to the world. In some instances, this fee was $15. With the release of *Meridian 59* and *Ultima Online*, bandwidth was no longer at a premium. This allowed the publishers of the games to charge monthly subscription fees instead of hourly rates for those who wanted to play the game. This proved to be a much more successful model, and *Ultima Online* quickly grew to 100,000+ subscriptions, a milestone for MMOs at this point in time. In 1999 NCSoft's *Lineage* broke the 1-million subscriber mark, another milestone in online games.

The 1990s also saw the release of many 3-D social worlds, where the goal was not to kill dragons but to interact with other players to complete common tasks. Alpha Worlds was an early example of an online social world, although it only could support 12 avatars simultaneously. The market for online social worlds is still unclear and many publishers continue to focus on dungeons and dragons–themed games. In the late 1990s AlphaWorld eventually morphed into Active Worlds, a virtual world platform provider still in business today.

Early 2000 continued to see the release of many MMORPG games, but the genre was slowly starting to expand out of the dungeons and dragons–themed games and into other genres. More social virtual worlds also began appearing, but it was not until 2003 when Second Life and There arrived that the social virtual worlds began picking up momentum.

Second Life, currently the most popular online social virtual world, is a world where users inhabit the environment via avatars and not only interact with one another, but actually build the world in real time. This is something that made the world very unique. Linden Labs, the creator of Second Life, created Second Life as more of a platform than a game-world, and provided the users the tools necessary to populate the world with a dizzying variety of content. Another unique aspect of Second Life is that it is one massive world. Most other virtual worlds utilize a technique called "shards," where many mirror images of a world exist, each supporting several thousand users. When one shard reaches capacity, a second shard is opened that can support an equal number of users. As the virtual world grows, more shards come online. Second Life chose not to shard the world, but rather to create a flexible, extensible architecture to support growth.

The other social virtual world, There, launched shortly after Second Life with similar fanfare. However, the inhabitants of There did not have the flexibility of those in Second Life in terms of creating objects and content within the world. After a couple of years, the military contracted the creators of There to develop and leverage the technology further for military use.

The final virtual world requiring mention is *World of Warcraft*, published by Blizzard Entertainment. *WoW* launched in 2004 and shattered all other virtual world sales figures, even those of *Lineage* which was hugely successful in Korea. Worldwide, *WoW* has an estimated 12 million subscribers, and is responsible for bringing the MMO genre to a mainstream audience.

10.3 TYPES OF VIRTUAL WORLDS

Beginning in the late 1990s and early 2000, it became clear that some of these virtual worlds could be loosely classified into different types or genres. These classifications are by no means official; virtual worlds are still in their infancy. Instead, this is a loose method of categorizing virtual worlds to assist the reader's understanding of the virtual world landscape. Some of these worlds reside in one category, but do contain characteristics that lend themselves to other categories.

The first, and likely the largest, category is the massively multiplayer online games, often referred to as MMOGs. This includes massively multiplayer online role-playing games (MMORPGs), massively multiplayer online first person shooters (MMOFPS), and other types of games where large numbers inhabit and interact in the same world. An important distinction is that these worlds are all games: the individuals that inhabit these worlds are considered players and have some sort of goal or objective the game imposes. Often these games are very complex and impose multiple goals on the player, each with varying degrees of complexity. This category could be partitioned into several subcategories, but for the purposes of this study that is not necessary because the focus lies in other virtual worlds.

An emerging category would be children's virtual worlds. Many of these worlds are very similar in nature to MMOs, but the world is clearly aimed at attracting children, whereas most MMOGs target older teens and up. One of the popular children's worlds is Disney's *Toontown*. This world is very similar to MMORPGs, but instead of taking on the role of a dwarf or elf and killing dragons, children take on the role of a Disney character and throw pies and other cartoon props at Disney villains. Another popular, although very different, children's world is *Club Penguin*. In this world, kids take on the role of a penguin, and can gather in one large lobby area before jumping out into smaller groups to participate in a variety of quick, fun mini-games such as racing, fishing, or just socializing. Disney has also launched *Pirates of the Caribbean*, a pirate-themed MMO that follows the adventures of Captain Jack and the characters from the movie of the same name.

The reason the children's worlds are worth a category of their own is to make the point that children are spending time in these worlds at a very young age. Before the proliferation of these worlds, children had no easy access to virtual worlds and generally would not see exposure until their late teen years or later. With Disney and other publishers creating virtual worlds aimed at children, we are seeing many more people inhabiting virtual worlds which may or may not lead to growing adoption overall.

The arrival and growth of Second Life have led to the emergence of social virtual worlds, where the inhabitants primarily use the environment to interact with one another. Unlike MMOGs, these worlds are not games: the world does not impose a goal or objective on the user. In these worlds, the user drives his or her experience. In a world like Second Life, this often involves users creating custom content using tools provided within the world.

The final category of virtual worlds focuses on worlds designed especially for education and training. Many worlds like Multiverse and Active Worlds are being used for education and training, but also for social purposes. These worlds tend to fit both categories, depending on implementation. A new virtual world called ProtoSphere appears to be the first virtual world aimed specifically at education and training. Not only is it a 3-D world inhabited by avatars, but the world contains integrated whiteboards, blogs, wikis, and application sharing. Companies are just starting to explore these environments as training platforms in a variety of settings.

Although not commercially available virtual worlds, Sun Microsystems and Lockheed Martin [A. Garga, private communication, November 2, 2008] are exploring the use of a virtual world as a part of the organization's intranet. Employees create avatars and navigate the 3-D space for training and collaboration opportunities. These recent efforts still appear to be in the experimental stages.

10.4 VIRTUAL WORLDS AND LEARNING

Understanding the virtual landscape provides a foundation to better understand the types of learning that take place within these virtual environments. The field of educational gaming or serious games has been slowly gaining momentum over the last decade with several books and articles authored on the positive outcomes of gaming, specifically as it pertains to learning. Some games present content that is historically accurate and can enlighten the player about historical events, but this writing will focus less on the content of games, and more on what players learn within the context of playing games.

Based on the literature dealing with game-based learning, several scholars agree that players are learning a wide range of skills that can help outside the context of a game. These skills include:

- Complex problem solving;
- Failure and persistence;
- Multitasking;
- Pattern recognition;
- Collaboration.

A characteristic of a game is that it has a *win state*. Some games have a fairly simple win state, for instance, *Pac-Man*. Move around the game environment eating all the dots before a ghost touches you, and you win. Other games have a much more difficult win state to achieve, for example, *Civilization*. In *Civilization*, the player chooses a culture to represent and spends many hours advancing his or her chosen culture. The win state comes when the player's culture has taken over the game world (the European continent). To get to the win state, the player faces a very large challenge or problem. In *Civilization*, it takes a combination of scientific advances, diplomacy, religion, and warfare strategies to win. No singular path to a win state exists. This provides a canvas for the player to manipulate the game world in a cause-effect fashion, to probe the game in order to solve the problem of world dominion.

This also leads to the idea of failure and persistence. In nearly every complex game today, it takes several failures in order to finally succeed and enter the win state. Players are often willing to spend more than 40 hours learning how to play a game correctly. For example, in *Civilization*, you quickly learn that some of the cultures that the player faces are gifted in specific areas. Some are advanced in military strategy and technology, but in order to find this out, it often takes one of the player's convoys getting wiped out. Players probe the game world, strategize, often fail, and then restrategize in order to win. The idea of failure and persistence can be equally important outside of a game context and, for example, in the workforce.

Players also learn how to multitask. In many games, the player needs to manage not only the goal of winning the game, but also many subgoals that allow for varying levels of progression. This is very prevalent in real-time strategy (RTS) games, where a player needs to manage his or her own resources, constantly using resources to grow the player's power in the environment. While the player is managing and allocating resources, either the computer or other players are doing the same thing within the same game world, so each player is also charged with monitoring the opponents.

Pattern recognition is also important in games and can help a player get to the win state much sooner. This takes place a great deal in puzzle and shooter games, where the player can gain an upper hand if he or she identifies patterns with the artificial intelligence driving the game. Outside of video games, the same can be said about athletic games like football. If a defensive lineman notices the quarterback raises his right heel a second before the ball is snapped, that lineman is at a great advantage.

Finally, online games show great promise in terms of collaboration. This is especially true in MMOGs, where it often takes more than 40 individuals to group together in order to attain a common goal. Most of these games are structured to encourage the player to work with other players. In some games it may only take one other person to assist in achieving a goal, whereas in other games it may take 100 other players. Research is beginning to emerge regarding the skills players are learning when participating in these group-oriented events. Specifically, literature points to:

- Collaborative problem solving;
- Distributed collaboration and communication skills;
- Conflict resolution;
- Leadership skill building.

10.5 THE WIKIPEDIA PHENOMENON

As of this writing, the English version of Wikipedia contains more than 3 million articles, created and edited by over 1 million users. For those not familiar with the process in which Wikipedia operates at a high level, it proceeds as follows:

- Anyone with an Internet connection and a Web browser can go to www.wikipedia.org and create an account at the site.
- Once a user has an account and logs in, he or she can begin to either create new content in the form of a stub (small collection of text around a specific topic) or edit existing information within Wikipedia all using an online collaborative editing platform called a wiki.
- Upon completing the edit, the user can then save the page and it will immediately be viewable by other users exploring Wikipedia.

When Wikipedia launched in January 2001, many onlookers felt that the project was doomed to failure: "How on earth will we know if this information is accurate? If anyone can create and edit pages, what will stop users with personal agendas from polluting the collection of information?" To combat these concerns, a variety of mechanisms were put in place.

The team behind Wikipedia created the Five Pillars (http://en.wikipedia.org/wiki/Wikipedia:Five_pillars); a high-level overview of the guidelines and policies that drive the site:

1. Wikipedia is an encyclopedia.
2. Wikipedia always maintains a neutral point of view.
3. Wikipedia is free content.
4. Wikipedia has no code of conduct.
5. Wikipedia does not have firm rules.

This helps to understand the goals of Wikipedia, but it still does not clarify how the content will be managed. What will stop someone from creating racist content or materials that are sexually explicit or are substantially inaccurate?

No one person could possibly handle this chore. Remember, over 3 million pages of content exist inside Wikipedia and that number continues to grow. Who is responsible for ensuring neutrality and accuracy? Anyone with an Internet connection and a browser can volunteer to be responsible for these duties, including you and me. Thousands of individuals around the world participate as editors in Wikipedia. Editors can choose a variety of roles, including quality control, vandalism patrol, and peer review.

"The Wikipedia community is largely self-organizing, so that anyone may build a reputation as a competent editor and become involved in any role they may choose, subject to peer approval. Individuals often will choose to become involved in specialized tasks, such as reviewing articles at others' request, watching current edits for vandalism, watching newly created articles for quality control purposes, or similar roles. Editors who find that editorial administrator responsibility would benefit their ability to help the community may ask their peers in the community for agreement to undertake such roles, a structure which enforces meritocracy and communal standards of editorship and conduct. At present around a 75–80% approval rating after enquiry is considered the requirement for such a role, a standard which tends to ensure a high level of experience, trust and familiarity across a broad front of projects within Wikipedia" [7].

For those not familiar with the concept of Web 2.0, it is often characterized as a web of participation. Wikipedia has leveraged this participation successfully, with volunteers from around the world checking content daily and taking the appropriate actions when errors or misinformation is found. Recent studies are beginning to show this process not only works, but it actually works *better* than some traditional publishing approaches in terms of accuracy (cites). Wikipedia was found to be more accurate than *Encyclopaedia Britannica*, and articles that deal with rapidly changing fields such as information technology can maintain accuracy over time within Wikipedia, where traditional publishing models such as journals and books have a latency time that occasionally makes the final publication irrelevant shortly after press.

This model of community self-organization and participation is a model that could also work to create an ad hoc community of analysts. These analysts could come together naturally in a virtual world to provide valuable insight on emerging situations. Similar to Wikipedia, community members could recognize one another using some sort of rating system. This allows for knowledgeable volunteers within the system to take on more responsibility and be given more tools to facilitate and capture knowledge within the system.

Imagine a situation in which an emerging environmental disaster threatens a city or region (e.g., the Hurricane Katrina disaster in New Orleans). Dynamic evolving information related to logistics, weather phenomena, environmental impacts, health conditions, medical phenomena, and others could be shared via a Wikipedia designed especially for such a specialized event or threat. Local residents might provide information about specialized groups (e.g., patients in assisted-care facilities) or seek specialized knowledge from local farmers or builders with particular knowledge about local structures. Distributed domain experts could begin a collaboration to define how to assess the evolving situation and assist in developing plans for mitigation.

10.6 VIRTUAL WORLD COLLABORATION TOOLS AND PLATFORMS

Many virtual world tools and platforms exist today. Once Second Life took off as a platform for collaboration and education, increasingly more virtual world tools were being developed to expand on this concept. This section provides an overview of tools and platforms that either are available now or are in the final phases of production.

10.6.1 Second Life

Second Life is arguably the most popular virtual world tool in use today, outside of the MMORPG worlds. Second Life was launched in June 2003 with little fanfare. Early users seemed unsure of what to make of Second Life. Is it a game? Is it a 3-D chat room? Is it some sort of social experiment? Early adopters began building and scripting objects in the world utilizing the 3-D modeling tools that are packaged within the client software. The platform was also somewhat buggy, with unannounced server restarts and patches occurring frequently.

In 2006, the mainstream media started to take notice of Second Life and several articles began appearing in newspapers and magazines around the United States. At this point, some users had created their own businesses within Second Life and were making a living in the virtual world. A landslide of media attention followed that still persists today, although some of the media attention has centered on the negative aspects of Second Life, such as the adult-themed areas.

Second Life is unique in two ways:

1. Users are provided the tools in the client software to create nearly anything imaginable.
2. Second Life contains a unique economy where users can use an exchange system to convert real U.S. dollars to Linden dollars (Second Life currency), as well as convert Linden dollars to real U.S. dollars.

In one sense, Second Life is similar to a virtual Lego Universe, where users can create things in the world by combining 3-D objects and using a scripting language. This appears to be a driving factor of why so many organizations have adopted Second Life for a variety of projects, including advertising, training, education, and product development. Organizations such as IBM, Dell, Sun Microsystems, BMW, the NBA, and many others have (or had at one point) a presence inside Second Life. Over 100 universities around the world are also experimenting with Second Life, using it as a medium for online collaboration and online courses (see Figure 10.1).

Figure 10.1 A screenshot from a conference held within Second Life.

Second Life provides a safe "first step" into virtual worlds for many individuals and organizations. For about $2,500 one can purchase a 3-D island within Second Life and a year of maintenance fees. With this in mind, Second Life is very similar to an Internet service provider.

Second Life does have a few disadvantages, particularly when it comes to protecting an organization's data. Because Second Life is like an ISP, all the servers reside in Linden Lab's server farms. IBM is working with Second Life to create an environment where organizations can host their own installations of Second Life, but still connect to the main grid (the primary servers that comprise the majority of Second Life). Second Life can also be somewhat difficult to learn, particularly the Linden Scripting Language. A comparison of traditional Internet Service Providers (ISP) with Lindon Lab services is shown in Table 10.1.

Table 10.1

Comparison of Traditional Internet Service Providers with Linden Labs

Traditional ISP	*Linden Labs (Owners of Second Life)*
Provides Web server space	Provides 3-D server space
Register a domain name	Register an island name
Create content using HTML files	Create content using 3-D objects
Interactivity via PHP, ASP, ColdFusion	Interactivity via Linden Scripting Language (LSL)

10.6.2 OLIVE by Forterra

Around the same time Second Life was released, a similar virtual world called *There* arrived. There was not as robust as Second Life in terms of user-generated content, but it was similar to Second Life in that it was not a game but more of a social virtual world where users could interact in 3-D, engage in mini-games like racing, and create small quests for other users. The world did feature minimal user generated content: utilizing photo manipulation programs like Photoshop, users could create things like clothing for other users to purchase via There bucs, the currency of There.

After a public release in October 2003, in 2005 There Inc. restructured into two companies: Forterra Systems, which focuses on government contracts, and Makena Technologies, which maintains the commercial virtual world of There. After the split, Forterra began working with the U.S. military, creating virtual worlds for military training and expanding on the capabilities of the technology platform. This eventually led to their OLIVE platform, which stands for On-Line Interactive Virtual Environment. This platform is currently used in government, corporate, defense, medical, and educational organizations. Forterra offers both hosted installations of OLIVE as well as installs at client locations. Forterra can also assist in custom content creation within the OLIVE environment, or clients can create and import their own content.

In January 2008, the IEEE recognized the OLIVE platform as one of six technological winners, the only virtual world platform to receive an award [5]. An example of a screenshot from OLIVE is shown in Figure 10.2. The Serious Games Institute at Coventry University also chose OLIVE as the major platform that the university program will utilize to deploy virtual worlds and game-based learning initiatives. In 2007, the Defense Department's Joint Advanced Distributed

Figure 10.2 A screenshot from OLIVE, taken from http://www.forterrainc.com.

Learning Laboratory selected Forterra to research methods for enabling SCORM-compliant (Sharable Content Object Reference Model) content to be utilized in virtual worlds, which places Forterra in a great position to help define how virtual worlds handle SCORM-compliant learning content.

10.6.3 ProtoSphere by Proton Media

ProtoSphere is a rapidly growing virtual world developed by Proton Media from suburban Philadelphia. ProtoSphere is unique when compared to both Second Life and the OLIVE platform. Unlike Second Life, ProtoSphere is designed specifically for education and training. Unlike OLIVE, ProtoSphere can be an "out-of-the-box" solution for some organizations. ProtoSphere more closely resembles OLIVE than Second Life, but ProtoSphere includes an additional social networking layer that enables users to connect with other users in ways that resemble how people connect on social Web sites such as Facebook, MySpace, and LinkedIn. This can be an extremely powerful feature when implementing ProtoSphere in a massive organization, with locations spread across the country or the world. Employees have the opportunity to connect with one another not only through work-related projects, but also to come together around similar interests and passions. This greatly increases the organization's potential for collaboration and innovation.

In addition to the social network component of the application, ProtoSphere has a rich feature set including:

- Voice over IP communications (VoIP);
- Collaborative whiteboards;
- Decision tree interactions;
- Text-based chat and buddy lists;
- Blogs;
- File sharing;
- Application sharing; and
- Team or group creation and custom permissions.

ProtoSphere does not include an integrated 3-D modeling toolset in its client software like Second Life. Organizations that leverage ProtoSphere can contract Proton Media for custom creation, contract a third-party developer, or work with internal staff to develop custom content. ProtoSphere leverages 3-D Studio Max for 3-D asset creation and utilizes the LUA programming language, a game

Figure 10.3 A screenshot of ProtoSphere.

industry standard, for scripting interactions within the environment. Finally, XML is used to capture metadata around objects and interactions that occur in the environment. Organizations have the option of purchasing and installing ProtoSphere internally, or working with Proton Media for an external hosting solution. An example of a ProtoSphere screenshot is shown in Figure 10.3.

10.7 LESSONS FROM ONLINE, MULTIPLAYER GAMING COMMUNITIES

We have conducted several experiments at The Pennsylvania State University involving the use of virtual world technologies in class projects using computer-savvy undergraduate students. We describe the results here, since we believe that they are representative of issues experienced by any first-time or casual user of the technology. The following are some of our initial lessons learned:

- *The need for clear, goal directed activities:* Students do not seem to do well with basic, "exploration" assignments in environments like Second Life. However, when given a very clear activity, such as creating an interactive sign in Second Life, students do very well and seem to enjoy the creation aspect of the activity. This has implications for virtual team collaboration.

- *Implementation and scalability challenge:* It is necessary to conduct design and development very early and to test simulations in worlds such as Second Life very often before rollout to make sure permissions are 100% correct and students have the ability to do what is necessary and only what is necessary. We have had instances where students have vandalized other students' Second Life work. This is what we have found works well:
    - Create a batch of avatars that is managed by a single person or spreadsheet.
    - Create all groups and permissions before experiments begin. Do everything possible up-front so students/participants do not have to (create avatars, groups, database connections).
    - Many scalability problems, such as avatars can only be associated with 25 groups in Second Life and Second Life servers only support 40–50 simultaneous users, forced patch updates from Linden Labs.
- *Unexpected learning curve:* Students find environments like Second Life and ProtoSphere somewhat confusing at first, particularly in Second Life, where things take a great deal of time to load, the environment is buggy, and so forth. Student users made frequent comments like: "The interface is getting in my way" and "this platform is broken." However, these comments are from 18- to 21-year-olds who play video games, so they are used to polished, game-like environments. In order to use the environment effectively, users should be provided with plenty of up-front time to learn the environment, and even go so far as to create some sort of beginner activity to help them do this (for example, Second Life scavenger hunt).
- *Avatar impact on sense of presence:* We found the avatar in a virtual world can do a great deal to establish a sense of presence and social connections with other individuals. The Pennsylvania State University's College of Information Sciences and Technology (IST) has staged contests in Second Life where we have faculty and students from around the state participate. Such contests support good community building and help people feel connected.
- *Immersive qualities:* Simply putting users in a virtual world with rich 3-D graphics *does not* make it immersive. A balance is required between a sense of immersion and a sense of engagement in the virtual world. If participants can be drawn in with an engaging, goal-directed activity, the students report feeling more immersed in the world.

- *Difficulties with assessment:* Virtual worlds, to this point, have not been created with tools to assess user/avatar actions. Working with faculty in Second Life, it is very difficult to find the right method to assess student behavior in the world. It is difficult to implement tracking mechanisms and other methods to assess activities in the world.

- *Use outside of class:* In our experiments we found that less than 50% of students had heard of Second Life before arriving at IST. However, once Second Life is introduced in classes, we find that a small number of students latch on to the environment and begin spending several hours a week in Second Life outside of classwork.

- *Customization:* Second Life showed everyone how powerful an environment can be if you allow user-generated content. Most virtual worlds (outside of MMOs/subscription model games like *World of Warcraft*) are incorporating this on some level and others are making this the key selling point for emerging worlds.

- *Distributed collaboration:* Many similarities are drawn from the way self-organizing groups operate in games like *World of Warcraft* with the way distributed teams work in software development. In *World of Warcraft*, groups of people gather together, each with varying skill sets, to achieve common goals. It takes an extremely high level of organization, strategy, and execution to succeed. This is similar to a distributed team working on a software development project. The major difference is that in a game like *World of Warcraft*, everything is within the context of the game, from start to finish. With team collaboration in a virtual world, the focus is on using the platform for interaction. Many of the activities, most importantly, the final outcome, take place outside the virtual world.

- *Emergence of new environments:* New environments are emerging and include features such as those listed here:

    - Better workflow: virtual world environments are getting better and better at incorporating existing forms of content and industry standard tools. Examples include creating content in 3-D Studio Max and incorporating it into Multiverse worlds and using XML-RPC to move data in/out of Second Life to the external Web. Another example is the use of Flash with applications like a Smart Fox Server to build browser-based virtual worlds.

    - Web integration: children's worlds like *Whyville*, *Webkinz,* and *Club Penguin* are showing us the importance of putting worlds on the Web. It is becoming much easier for people to find and adopt. Hence, there are lower barriers to entry. Raph Koster's Metaplace looks promising and will take this a step further.

- Data fusion: worlds like Multiverse and the OLIVE engine are getting better at accepting incoming data streams to trigger events in-world, much like *Madden NFL* taking live weather feeds into the game and *NBA Live's "Player DNA"* project where live data impacts the environment in a realistic way.

10.8 SUMMARY

Rapid advances in virtual world technologies and commercial distribution of virtual world tools allow large groups of people to collaborate in an ad hoc manner. This can be performed asynchronously using a concept such as Wikipedia, in which self-proclaimed experts post information about a wide variety of subject areas to create a worldwide, world-developed encyclopedia. Conversely, environments such as Second Life or OLIVE allow participants to synchronously interact in a common, shared virtual world via avatars. A new generation of Web-literate analysts is thoroughly familiar and comfortable with such environments. Indeed, the advent of online children's worlds such as Disney's *Toontown* makes such interactions very natural. It remains to be seen how these new technologies and user orientations will evolve to support distributed analysis of complex problems such as homeland security, national defense, or crisis management and response.

References

[1] Bell, M., "Definition and Taxonomy of Virtual Worlds," http://www.slideshare.net/storygeek/definition-of-virtual-worlds.

[2] Koster, R., "Online World Timeline," February 20, 2002, http://www.raphkoster.com/gaming/mudtimeline.shtml.

[3] Damer, B., "Origins and Evolution of Social Virtual Worlds," 2007, http://www.vwtimeline.org/presentations/pub-launch-vwtimeline/linden_files/frame.htm.

[4] Gordon, E., "History of Virtual Worlds," 2007, http://www.avatarplanet.com/history.php.

[5] Kushner, D., "Winner: Software—Make Your Very Own Virtual World With OIVE," *IEEE Spectrum*, Vol. 45, No. 1, 2008, pp. 45–47.

[6] Dibbell, J., "A Rape in Cyberspace or How an Evil Clown, a Haitian Trickster Spirit, Two Wizards, and a Cast of Dozens Turned a Database into a Society," *The Village Voice*, 1993, pp. 36–42.

[7] http://en.wikipedia.org/wiki/Wikipedia:About#Contributing_to_Wikipedia.

Chapter 11

Information Markets and Related Forms of Collective Processing

This chapter introduces the concept of crowd-based sense-making—how to access knowledge by using a large group or crowd of people. The focus involves the concept of information markets in which information is treated as a commodity and market-based interactions are used to determine the true consensus of the large group.

11.1 CROWDSOURCING OF ANALYSIS

Chapter 1 introduced the concept of human-centered information fusion with new roles of humans throughout the fusion process. As indicated in Chapter 1, the traditional role of the analyst in information fusion systems had been one of a relatively passive observer to an evolving situation display (albeit with interaction to control the display and perform database retrievals).

The right side of Figure 11.1 shows the concept of using multiple, geographically distributed analysts interacting in a collaborative manner to assess a situation, perform analysis, or achieve consensus regarding a situation or problem. Traditionally, this type of interaction had been limited to a few, colocated analysts in a single room (e.g., a data analysis center or a crisis management center). However, rapid changes in communications and collaboration technology provide the opportunity for geographically distributed users to collaborate for group or "crowd-based" analysis. Technologies such as large-scale online multiplayer

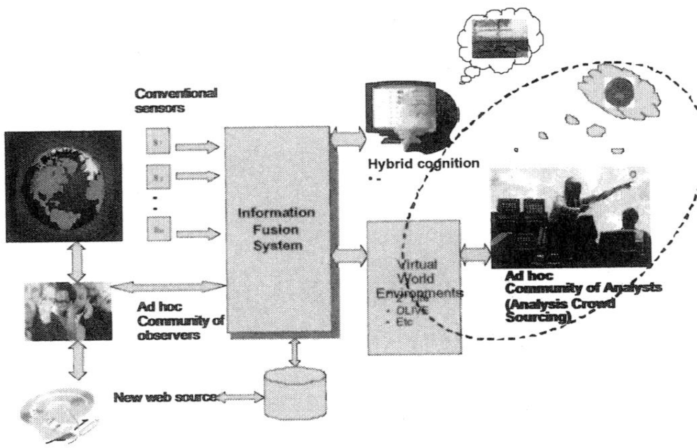

Figure 11.1 Concept of crowdsourcing of analysis.

games show the potential for large groups of people to interact for a common purpose (in this case to play a game) and to pit their collective expertise against a common problem.

This chapter and Chapter 12 explore the use of Web-based collaborative technologies for implementing the concept of analytical crowdsourcing: the use of an ad hoc distributed team of participants to collaboratively address a problem. In particular, this chapter explores the use of prediction markets and related methods to access group or crowd knowledge, while Chapter 12 explores the use of virtual world technologies for interactive, distributed collaboration.

11.2 THE WISDOM OF CROWDS

Only four years after the release of James Surowiecki's book *The Wisdom of Crowds* [1], it has become an article of Internet faith that groups of people can give better answers than individuals will. The mechanisms whereby groups can do collective sense-making, however, are still not fully understood. Even so, what is known tends to support the faith that many people place in these modes of information processing.

The purest form of collective information processing is probably the prediction market. Just as financial markets synthesize multiple points of view on the future prospects of a company into a price for stocks or other instruments, prediction markets gather multiple informed and interested participants who buy positions on the likelihood of a future event. The concept is analogous to the Las Vegas sports book, which prices options for bettors' futures contracts (wagers) on the outcome of an upcoming game or match. The point spread is a pricing mechanism to divide the betting population into relatively equal segments, minimizing the house's exposure to an asymmetric payout.

In addition to the formidable track record compiled by the Iowa Electronic Markets (http://www.biz.uiowa.edu/iem/), an early private-sector success story involved an experimental prediction market at Hewlett Packard in the late 1990s. Employees bought futures contracts related to both specific product sales and overall corporate profitability. Information was consolidated about major accounts (which might involve interaction with multiple business units), unit pricing strategies, and overall market dynamics. Individual trades were visible to other traders, but anonymized. The academics who oversaw the experiment found that markets outperformed traditional forecasting, sometimes with an uncanny degree of accuracy [2]. Those preliminary results were widely reported among consultants and trend-watchers, helping fuel a larger interest in the practice. In general, however, after three years the improvements were not consistently better than previous methods, and a new model has since been devised. Google has since adopted internal prediction markets but has told outsiders few details of their operation [3].

11.3 HOW DO CROWDS EXPRESS WISDOM?

How do crowds express a collective wisdom? Several mechanisms have been used:

- *Voting methods:* Voting involves the collection of official or ad hoc feedback regarding a situation, decision, or process. This may entail formal processes such as used in politics or formal meetings, or unofficial feedback provided via product reviews, Digg, or similar feedback ("Was this review helpful?"). All of these actions are voluntary and unsolicited, making statistical significance a moot point. Self-selection bias is a major concern in informal voting; a real person may not even be clicking, a development that prompted the invention of the CAPTCHA, the wavy letters that seek to determine if an actual human is behind certain Internet transactions.[1] Voting is commonly thought of in terms of a democratic process involving a single vote for each participant and a majority rules concept. However, there are a wide variety of methods ranging from unweighted voting using a simple majority or plurality rule to weighted voting schemes. Hall and McMullen [4] have provided an overview of voting techniques. Miller and Hall [5] have developed a method involving the use of automated monitors or critics who select the optimal voting population based on the judged expertise of the voting (sensors or information sources).

- *Betting:* Betting seeks to obtain accurate feedback using monetary concepts to allow individuals to quantify the extent of their support for a selection or decision [6]. In some cases this involves real money where people place real bets (as at Iowa Electronic Markets (IEM), which predicts elections) or *imagined* money (at Hollywood Stock Exchange (HSX), which specializes in movie grosses and other cinematic developments). In either case, people are asked to put their currency behind their convictions. Given the right kind of topic and the right kind of crowd, this process can be extremely powerful, albeit with constrained questions. This type of approach may also be used to help solicit a priori probabilities ("priors") for Bayesian inference processes [7]. A betting approach seems to help people be more decisive in their assertion of the likelihood of an event or causal relationship, compared to the abstract use of subjective probabilities.

- *Surveys:* We are familiar with the concept of surveys [8] due in part to their constant use in predicting election results and obtaining feedback for commercial products. These are often constructed with elaborate statistical tools and focused on carefully focused questions. Interaction among respondents is usually low, making surveys useful in collecting independent opinions. Tools such as Zoomerang make the construction and distribution of surveys via the Web very easy. However, the actual process of creating effective questions that do not "lead" a respondent to a presumed answer and the evaluation of the accuracy of a surveyed population (namely, how representative the surveyed population is to the sought-after population) can be very challenging. Failure to construct accurate questions, inappropriate selection of a surveyed subpopulation, and naïve statistical interpretation of results can lead to significant failures in a survey process.

- *Convened feedback:* This catch-all includes tagging, blogs and comments, message boards, trackbacks, wikis, and similar vehicles. Once again, the action is voluntary, but the field of play is unconstrained. Compared to the

1 According to the inventing researchers from Carnegie Mellon University, the acronym stands for "Completely Automated Public Turing test to tell Computers and Humans Apart." See http://www.captcha.net/.

other three categories, convened feedback can contain substantial noise, but its free form allows topics to emerge from the group rather than from the pollster, market maker, or publisher. Convened feedback may also involve use of methods such as focus groups. An interesting example of such feedback is the Delicious social bookmarking Web site. This site allows people to share their favorite bookmarked Web sites and provides a mechanism for users to access the experiences of others to investigate interesting sites associated with a particular topic area. Thus, the site allows a collective mind to investigate and share links on the Web. Such techniques could be used by information analysts to share experiences involving favorite or useful information sources, types of analyses that assist in common problems, and methods to provide commentary on the reliability of sources and methods of analysis.

Note the difference in political predictions between polling and markets. Polls ask respondents, "Who is doing a better job?" "Who better represents your perspective?" and "Who do you intend to vote for?" Markets ask for a financial commitment to your projected winner, regardless of your personal feelings for or against a candidate. Like sports results, political markets work well because of their time-bound, binary outcomes. Other prediction markets with less discrete outcomes are harder to make liquid. That is, finding a "critical mass" of both buyers and sellers across a time horizon ("When will the number of U.S. troops in Iraq fall below *X* number?") is more difficult than in situations where the number of choices is finite and the time frame for determination is obvious.

11.4 WHAT KINDS OF QUESTIONS BEST LEND THEMSELVES TO GROUP WISDOM?

On this topic Surowieki [1] was direct: "Groups are only smart when there is a balance between the information that everyone in the group shares and the information that each of the members of the group holds privately." Conversely, "what happens when [a] bubble bursts is that the expectations converge" (pp. 255–256).

An excellent example of this effect can be found at Metafilter. In 2007, the question was posed, "What single book is the best introduction to your field (or specialization within your field) for laypeople?" Hundreds of people replied, in areas from homicide forensics to astrophysics. The results are priceless, a distillation of centuries of experience into a modest library (http://ask.metafilter.com/71101/What-single-book-is-the-best-introduction-to-your-field-or-specialization-within-your-field-for-laypeople).

Cass Sunstein, a University of Chicago law professor, agreed in his book *Infotopia* [9]. He stated that, "This is the most fundamental limitation of prediction markets: They cannot work well unless investors have dispersed information that can be aggregated" [9, pp. 136–137]. Elsewhere in a blog posted he noted that in an informal experiment with University of California law professors, the crowd came extremely close to the weight of the horse that won the Kentucky Derby, did "pretty badly" on the number of lines in Shakespeare's *Antigone*, and performed "horrendously" when asked the number of Supreme Court invalidations of state and federal laws. He speculates that markets employ some self-selection bias: "participants have strong incentives to be right, and won't participate unless they think they have something to gain."

The best questions for prediction markets, then, involve issues about which people have formed independent judgments and on which they are willing to stake a financial and/or reputational investment. It may be that the topics cannot be too close to one's professional interests, as the presence of bubbles in financial markets would suggest on one hand, and in line with the accuracy of the HSX Oscar predictions on the other.

11.5 WHERE IS ERROR INTRODUCED?

The French political philosopher Condorcet (1743–1794) originally formulated the jury theorem that explains the wisdom of groups of people, when each individual is more than 50% likely to be right. Bad things happen when people are less than 50% likely to be right, however, and crowds then amplify error. This concept extends to the fusion of multisource/multisensor data. Failure to accurately characterize the reliability of sources in the context of a real observing environment can lead to fusion results that are significantly worse than the information reported by a single sensor or source [5].

Numerous experiments have shown that group averages suffer when participants start listening to outside authorities or to each other. What Sunstein called "dispersed information" and what Surowiecki contrasts to mob behavior—independence—is more and more difficult to find. Many of the start-up companies in idea markets include chat features—they are, after all, often social networking plays, making for yet another category of echo chamber.

Another kind of error comes when predictions ignore randomness. Particularly in thickly traded markets with many actors, the complexity of a given market can expose participants to phenomena for which there is no logical explanation, even though many will be offered. As Nassim Nicholas Taleb pointed out in *The Black Swan* [10], newswire reports on market movement routinely and fallaciously link events and price changes. It is not uncommon to see the equivalent of both "Dow falls on higher oil prices" and "Dow falls on lower oil prices" during the same day.

11.6 VARIETIES OF MARKET EXPERIENCE

The following are just some of many businesses seeking to monetize prediction markets:

- Newsfutures (http://us.newsfutures.com/home/home.html) makes a business-to-business play, building internal prediction markets for the likes of Eli Lilly, the Department of Defense, and Yahoo!.
- Spigit (http://www.spigit.com/) sells an enterprise software to support internal innovation and external customer interaction. Communities are formed to collect and evaluate new ideas.
- Intrade (http://www.intrade.com/) is an Irish firm that trades in real money—with a play money sandbox—applied to questions in politics, business (predictions on market shares are common), entertainment, and other areas. The business model is built on small transaction fees on every trade.
- Hubdub, from Edinburgh, trades in play money but prominently features leaderboards, which intensify user involvement. Topics under discussion are limited only by users' imaginations and curiosity, as any member can propose a question. A recent leader, named Orlin, has done well on European football but also advanced wide-ranging predictions, including one regarding the Higgs boson being discovered by the large hadron collider within a year. He or she has made nearly 6,000 predictions.

Apart from social networking plays and predictions, seemingly trivial commitments to intellectual positions work elsewhere. Cass Sunstein's more recent book, called *Nudge* [11], was coauthored with the Chicago behavioral economist Richard Thaler. It points to the value of commitment for such personal behaviors as weight loss or project fulfillment. For example, a Ph.D. candidate, already hired as a lecturer at a substantial discount from an assistant professor's salary, was behind on his dissertation. Thaler made him write a $100 check at the beginning of every month a chapter was due. If the chapter came in on time, the check was ripped up. If the work came in late, the $100 went into a fund for a party to which the candidate would not be invited. The incentive worked, notwithstanding the fact that $400 or $500 was a tiny portion of the salary differential at stake. A Yale economics professor who lost weight under a similar game has cofounded stickK.com, an ad-funded online business designed to institutionalize similar "Commitment Contracts."

11.7 FUTURE DIRECTIONS

It is clear that crowds can in fact be smart when the members do not listen to each other too closely. It is also clear that financial and/or reputational investment is connected to both good predictions and fulfilled commitments. Several other issues are less obvious. Is there a novelty effect with prediction markets? Will clever people and/or software devise ways to game the system, similar to short-selling in finance or sniping on eBay [12]? What do prediction bubbles look like, and what are their implications? When are crowds good at answering questions and when, if ever, are they good at posing them? (Note that on most markets, individuals can ask questions, not groups.) Can we reliably predict whether a given group will predict wisely?

At a larger level, how do online information markets relate to older forms of group expression, particularly voting? The United States' filtration of a state's individual votes through the winner-take-all Electoral College is already controversial (only Maine and Nebraska currently allot their votes proportionately), and so-called National Popular Vote legislation is passed or pending in states with 274 electoral votes, enough to overturn the current process. Will some form of prediction market or other crowd wisdom accelerate or obviate this potential change?

Any process that can, under the right circumstances, deliver such powerful results will surely have unintended consequences. The controversy over John Poindexter's Futures Markets Applied to Prediction (FutureMAP) program, which was cancelled by DARPA in July 2003, will certainly not be the last of the tricky issues revolving around this class of tools.

11.8 FUTUREMAP: A BRIEF CASE STUDY

The U.S. Department of Defense, through its Defense Advanced Research Projects Agency (DARPA), pursues research and development in both basic and applied science. The Internet began under DARPA auspices; more recently, the agency's Grand Challenge advanced the state of unmanned vehicular navigation and performance. After the September 11, 2001, attacks, the limits of conventional intelligence gathering in the face of an unconventional terror threat spurred efforts to develop new tools of information analysis. According to DARPA, such "analysis often requires independent contributions by experts in a wide variety of fields, with the resulting difficulty of combining the various opinions into one assessment. Market-based techniques provide a tool for producing these assessments." In short, DARPA was using its mandate to explore the science of information markets to address a topic area—the future of the Middle East—with pressing needs [13–16]. An economist from George Mason University named Robin Hanson, who formerly worked for both NASA and Lockheed, was named to help lead the effort, called Policy Analysis Market, or PAM. The lead vendor was a company called Net Exchange, which provided a technology called a "combinatorial automated market maker" (CAMM).

Contrary to later characterizations of the project as a "terror market," PAM was more centrally concerned with broad-based economic, political, technological, and military trends [14]. The project description made this point clear:

> Analysts often use prices from various markets as indicators of potential events. The use of petroleum futures contract prices by analysts of the Middle East is a classic example. The Policy Analysis Market (PAM) refines this approach by trading futures contracts that deal with underlying fundamentals of relevance to the Middle East. Initially, PAM will focus on the economic, civil, and military futures of Egypt, Jordan, Iran, Iraq, Israel, Saudi Arabia, Syria, and Turkey and the impact of U.S. involvement with each.

The "underlying fundamentals" were said to be "objective data and observable events" analogous to oil prices; some of this data as sourced from the widely respected Economist Intelligence Unit. Markets were chosen in part because information propagates widely and quickly: "the rapid reaction of markets to knowledge held by only a few participants may provide an early warning system to avoid surprise" [16]. As numerous projects focused on collaborative information-sharing have demonstrated, motivation and incentive present significant challenges in large organizations. To this end, "This price discovery process, with the prospect of profit and at pain of loss, is at the core of a market's predictive power."

As with any market, particularly for a new type of good, issues of pricing, liquidity, and transparency had to be addressed; this was likely a matter of concern to the organizers. Since real money was involved, such topics as tax liability, motivation, and market manipulation had to be addressed [16].

> Markets must also offer compensation that is ethically and legally satisfactory to all sectors involved, while remaining attractive enough to ensure full and continuous participation of individual parties. The markets must also be sufficiently robust to withstand manipulation.

> FutureMAP will bring together commercial, academic, and government performers to meet these challenges.

In practice, investments were capped at $100 to limit the exposure of the U.S. government and to minimize the potential reward for acting on sensitive (in multiple senses of the word) information.

In addition, the sensitivity of military and intelligence data in a market involving both civilian and military individuals with varying levels of security clearance complicated the issue:

> The DARPA FutureMAP program will identify the types of market-based mechanisms that are most suitable to aggregate information in the defense context, will develop information systems to manage the markets, and will measure the effectiveness of markets for several tasks. A market that addresses defense-related events may potentially aggregate information from both classified and unclassified sources. This poses the problem of extracting useful data from markets without compromising national security [16].

This issue later helped seal the project's demise. Several observers, including one of the senators who led opposition to the market, raised the potential for adversaries gaming the system to inject false or potentially dangerous perspectives into the system.

Hopes were high for the market as it was about to open. Three types of futures contracts were set to be offered on PAM [16] (quoting from an archive of the site):

1. Quarterly contracts based on data indices that track economic health, civil stability, military disposition, and U.S. economic and military involvement in Egypt, Iran, Iraq, Israel, Jordan, Saudi Arabia, Syria, and Turkey.
2. Quarterly contracts that track global economic and conflict indicators.
3. Specific possible events (e.g., U.S. recognition of Palestine in the first quarter of 2005).

The tool was portrayed in the terms more typically associated with a game than intelligence processing: "PAM will be active and accessible 24/7 and should prove as engaging as it is informative." But public reaction to one minor feature generated colorful and effective criticism. The primary opposition was led by

Democratic Senator Byron Dorgan of North Dakota, and Democratic Senator Ron Wyden of Oregon, who combined moral outrage and mockery when he stated that, "The idea of a federal betting parlor on atrocities and terrorism is ridiculous and it's grotesque." A day after their press conference, the project was terminated. Another factor in the public reaction to PAM was the fact that the project was linked to John Poindexter, who was already associated with a controversial data mining operation known as Total Information Awareness.

According to the DARPA director who oversaw the shutdown of the project, "FutureMAP was a small program that faced a number of daunting technical and market challenges, such as: Can the market survive and will people continue to participate when U.S. authorities use it to prevent terrorist attacks? Can futures markets be manipulated by adversaries? Reconsidering those challenges in light of the recent concerns surrounding the program, it became clear that it simply did not make sense to continue our participation in this effort." The colorful rhetoric, which the director characterized as "recent concerns," revolved around the prospect that "some individuals will make money by pondering the unthinkable," in the words of two scholars who wrote an analysis of the episode [16].

Putting aside the polarizing figure of John Poindexter and overlooking the reasonably benign project description, that outrage was likely generated by one small feature in the publicly released screen grabs of the market interface. Figure 11.2 is the image DARPA and Net Exchange released (enhanced from its ghosted version as a background).

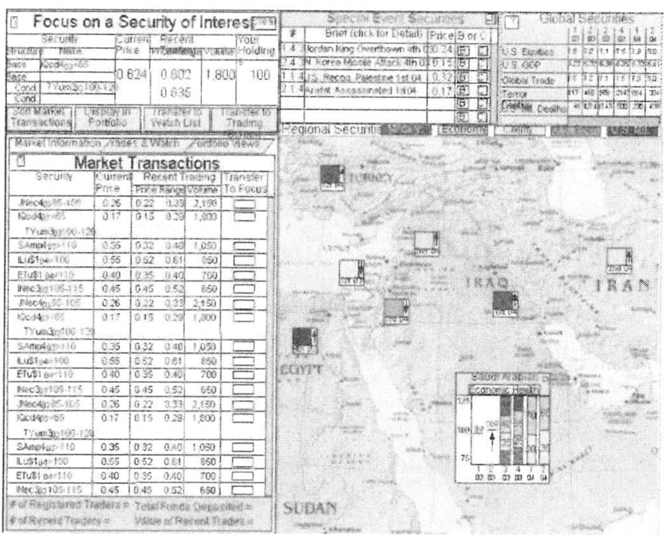

Figure 11.2 Screen capture of FutureMAP Net Exchange tool.

The problematic feature appears to have been a panel in the top center which focuses not on projected GDP growth or overall levels of political tension at a given border, rather, the "Special Event Securities" included a North Korean (nuclear?) missile attack and a political assassination. Had this one feature of the market been absent, or hidden behind a different tab, the reaction would likely have been unremarkable.

In the aftermath of the rapid dismantling of the effort, several things happened. Net Exchange attempted to sell the prediction market as a standalone product. Hanson performed an analysis of what he considered "uninformed" newspaper coverage [14]. Poindexter resigned from DARPA even though he was only tangentially involved in the "futures market for death," as then-Senator Hillary Clinton called it. The Hollywood Stock Exchange, Iowa Electronic Market, and other efforts built on James Surowiecki's book launched the wave of start-ups noted above. On Intrade, anyone who cares to can buy a futures contract on the capture of Osama Bin Laden within the next 8 months for about $12.

Despite the challenges of unfavorable publicity, prediction markets provide an opportunity for accessing "crowd wisdom" in the prediction of trends for intelligence analysis [17]. An overview of this potential is provided by Weigle [18] and Puong Fie Yeh [19].

11.9 SUMMARY

Rapid changes in Internet technology provide opportunities for dynamic collaboration of small to large groups for performing analysis. Methods ranging from online surveys to voting methods, betting techniques, and convened feedback via tagging, blogs, Wikipedias, and message boards make it increasingly easy to access distributed expertise. An emerging method for quantitatively accessing collective opinions related to technology forecasting, problem domain analysis, or causal analysis involves prediction markets. In this approach described in this chapter, participants use monetary concepts (including real and virtual money) to quantitatively express their opinions about activities, events, or situations. Prediction markets have been used to accurately forecast election results and estimate project schedules, product success, and other areas. The collection of such methods is promising for information analysis. We believe that such methods should be systematically explored for improving intelligence analysis (for applications ranging from business strategic planning to understanding crisis situations to forensic analysis of disasters). At this time the Web-based mechanisms exist for supporting dynamic collaboration, but the interaction implications are not well understood. Issues relating to how to characterize contributors' expertise, how to thwart electronic scamming and gaming (such as electronic auction sniping), how to combine group mind information with individual experts, and other issues remain unsolved. Nevertheless, this is a rich area of potential augmentation of information fusion and analysis.

References

[1] Surowiecki, J., *The Wisdom of Crowds,* New York: Random House, 2004.

[2] Chen, K.-Y., and C. R. Plott, "Information Aggregation Mechanisms: Concept, Design and Implementation for a Sales Forecasting Problem," Social Science Working Paper No. 1131, California Institute of Technology, Pasadena, CA, 2002.

[3] Cowgill, B., J. Wolfers, and E. Zitzewitz, "Using Prediction Markets to Track Information Flows: Evidence from Google," working paper, 2008.

[4] Hall, D., and S. A. McMullen, *Mathematical Techniques in Multisensor Data Fusion*, Norwood, MA: Artech House, 2004.

[5] Miller, D., and D. Hall, "The Use of Automated Critics to Improve the Fusion of Marginal Sensors for ATR and IFFN Applications," *Proceedings of the 1999 National Symposium on Sensor Data Fusion (NSSDF)*, Johns Hopkins University, APL, Baltimore, MD, May 1999.

[6] King, R., "Workers, Place Your Bets," *Business Week*, August 3, 2006.

[7] Pearle, J., *Probabilistic Reasoning in Intelligent Systems: Networks of Plausible Inference*, San Francisco, CA: Morgan Kaufmann, 1988.

[8] Groves, R., et al., *Survey Methodology*, New York: John Wiley and Sons, 2004.

[9] Sunstein, C. R., *Infotopia: How Many Minds Produce Knowledge*, Oxford, U.K.: Oxford University Press, 2006.

[10] Taleb, N., *The Black Swan: The Impact of the Highly Improbable*, New York: Random House, 2007.

[11] Thaler, R., and C. R. Sunstein, *Nudge: Improving Decisions About Health, Wealth, and Happiness,* New Haven, CT: Yale University Press, 2008.

[12] Ockenfels, A., and A. Roth, "The Timing of Bids in Internet Auctions: Market Design, Bidder Behavior and Artificial Agents," *AI Magazine*, Fall 2002, pp. 79–88.

[13] Hanson, R., "The Informed Press Favored the Policy Analysis Market," unpublished manuscript, August 8, 2005.

[14] Hanson, R., "The Policy Analysis Market: A Thwarted Experiment in the Use of Prediction Markets for Public Policy," *innovations,* Summer 2007, pp. 73–88.

[15] Kunz, B., "Prediction Markets Meet Wall Street," *BusinessWeek*, October 14, 2008.

[16] Looney, R., "DARPA's Policy Analysis Market for Intelligence: Outside the Box or Off the Wall?" *Strategic Insights*, Vol. 2, No. 9, September 2003, http://www.ccc.nps.navy.mil/si/sept03/terrorism.asp.

[17] Lyon, E., (ed.), *Theorizing Surveillance,* Cullompton, Devon, U.K.: Willan Publishing, 2006.

[18] Weigle, B. D., *Prediction Markets: Another Tool in the Intelligence Kitbag*, USAWC Strategy Report, U.S. Army War College, March 30, 2007.

[19] Yeh, P. F., "Using Prediction Markets to Enhance US Intelligence Capabilities," *Studies in Intelligence*, Vol., 50, No. 4, 2006.

Selected Bibliography

Berg, J., R. Forsythe, and T. Rietz, "What Makes Markets Predict Well? Evidence from the Iowa Electronic Markets," in W. Albers, et al. (eds.), *Understanding Strategic Interaction: Essays in Honor of Reinhard Selten*, New York: Springer, 1996, pp. 444–463.

Berg, J., F. Nelson, and T. Rietz, "Prediction Market Accuracy in the Long Run," *International Journal of Forecasting*, Vol. 24, 2008, pp. 283–298.

Erikson, R. S., and C. Wlezien, "Are Political Markets Really Superior to Polls as Election Predictors?" *Public Opinion Quarterly*, Vol. 72, No. 2, Summer 2008, pp. 190–215.

Pennock, D.M., et al., "The Power of Play: Efficiency and Forecast Accuracy in Web Market Games," NEC Research Institute working paper 2000-168, 2001.

Totty, M., "Business Solutions: How to Decide? Create a Market," *Wall Street Journal,* June 19, 2006.

Wolfers, J., and E. Zitzewitz, "Prediction Markets," *Journal of Economic Perspectives*, Vol. 18, No. 2, Spring 2004, pp. 107–126.

Chapter 12

Hybrid Cognition and Situation Awareness: Perspectives for the Future of Human-Centered Fusion

"Prediction is very difficult: especially if it's about the future"— Nils Bohr, Nobel laureate in physics

"What use could this company make of an electronic toy?"
—The President of Western Union, responding to Alexander Graham Bell's offer to Western Union of exclusive rights to the telephone for $100,000 in 1876

12.1 INTRODUCTION

Human-centered information fusion is emerging against a backdrop of political, economic, and technological change. The tools for, and objectives of, traditional fusion, meanwhile, are also shifting, making for an environment of extreme instability. The changes in these realms will place new demands on the analysts and decision-makers who must identify new types of threats and opportunities on a regular basis. In the near term, we expect technical systems and social patterns to coevolve, often in dramatic ways.

12.1.1 Reprising the Themes of Human-Centered Fusion

Data fusion systems have conventionally been premised on the use of physical sensors to observe physical targets to understand the physical landscape. For a wide variety of reasons, including rapid changes in information technology and shifts in political dynamics, a new human-centered view of information fusion is emerging to complement traditional methodologies. The key differences between the two perspectives include:

- *Observing the human landscape:* changing the domain of interest from the physical landscape to observing and characterizing the human landscape;
- *Soft sensors:* augmenting the use of physical sensors with humans acting as observers or "soft" sensors;
- *Hybrid computing:* using human-in-the-loop analysis in which humans use their visual and aural pattern recognition capabilities, along with semantic reasoning abilities, for analysis of complex data and situations;
- *Crowdsourcing of analysis:* using dynamic collaboration among multiple people to support analysis (e.g., via virtual world collaboration).

12.1.2 Dangers of Prediction

Human-centered fusion alters conventional intellectual trajectories: people have to revise how they think about all four of those domains. First, the *where* shifts from physical things to a combination of tangible (e.g., human gatherings) and intangible (electronic currency flows or attitudes) artifacts. Second, the *how* shifts from the use of strictly electromechanical devices to a combination of humans both as sensor platforms, by carrying mobile phones most obviously, and as information receptors and processors connected to digital networks. An example of the latter would be noting a large cash transaction at a fertilizer depot or recognizing a change in who makes mail or parcel deliveries. Third, human-centric fusion augments the *what* of information analysis, utilizing more human senses in an effort to utilize more human mental "bandwidth." Finally, the *who* shifts from individual contributors or strictly defined teams to a loosely organized community involving part-time and/or amateur and/or tangential professional resources.

Because human-centered fusion involves so many moving parts, making predictions about where all four vectors will or will not align is bound to be prone to error. This uncertainty is the product of both the rapid speed of technology change in the past 50 years and the involvement of people at new interfaces with the technologies. In particular, deterministic predictive models will fare extremely poorly in this domain. The magnitude of change is seldom proportional to the scale of the input (Facebook is one example), and the presence of essentially ubiquitous networking, whether technical or social, stresses the state of mathematical models. As we have noted, for example, statistical sampling techniques as yet do not work reliably in social network analysis.

12.2 REQUIREMENTS PULL FOR HCF AND SA

Demand for human-centered fusion and situational awareness is emerging against the backdrop of changes in the commercial, adversarial, military and defense, and demographic landscapes.

12.2.1 Commercial Forces

In the commercial realm, many facets of human-centered fusion are already in operation. BigChampagne, for example, tracks downloads of music on peer-to-peer networks to assess popularity and trends. In contrast to *Billboard* and SoundScan, which, respectively, measure music sales by surveying a sample of retailers or utilizing point-of-sale data, BigChampagne tracks user behavior at the level of the song rather than the album. Revenue, a measure of how much companies collect, reveals far less than download behavior, but even then no one has yet devised a measure of what people actually *listen* to (although Apple has some opt-in programs that move in this direction).

Many other aspects of the human landscape are the province of commercial data collection and analysis. Telecommunications providers measure multiple aspects of mobile phone usage, while researchers at MIT and Nokia developed tools for "reality mining" to build social network graphs on the basis of calling behavior [1]. To track consumer discussions of brands, Procter & Gamble recruited 600,000 online mothers with a desirable spread of social network behaviors such as credibility among their friends and willingness to try new products [2]. Private-sector information warfare is not unknown. The CEO of Whole Foods Market, John Mackey, posted derogatory statements under an assumed name on Yahoo! Finance message boards regarding a competing company he was trying to acquire; he also praised his own company in an apparent attempt to raise the stock price [3]. Even traditional market research, based on surveys, point-of-sale data, and external sources such as weather reports, can answer sophisticated questions about the inputs and outputs of consumer behavior.

At the same time, the commercially available sources of information that can contribute to human-centered fusion increase in complexity and richness every year.

- Google Earth, Maps, Traffic, and Street View provide images and information of obvious value.
- Consumer adoption of GPS, packaged in ever-smaller form factors, increases annually.
- Analytic software for tracking Web page visitors can reveal powerful patterns at low cost.
- Applications for mobile platforms, particularly the iPhone, allow dispersed individuals to collaborate or coordinate, as on a speed-trap notifier in which crowdsourced, geotagged locations are overlaid on a driver's current position.
- The Freesound project (http://www.freesound.org) provides geotagged audio files of ambient sounds from sites around the world.
- The amount of metadata contained in a single digital photograph, including geotagging, can be extremely useful (see Figure 12.1). Most people are unaware that some subset of this data exists in their photos.
- OpenStreetMap (http://www.openstreetmap.org/) applies open-source principles to mapmaking at a global scale.
- Social media tools allow anyone to broadcast thoughts, status, or images to a worldwide audience.
- Free tools allow the creation of video, 3-D, family trees, social network maps, and audio broadcasts.

12.2.2 Asymmetric Information Warfare

Given the capability of these tools, many groups are using them for criminal, political, ideological, and disruptive purposes. In less than 12 months, the Mumbai attacks of 2008 and the Iranian election demonstrations of 2009 demonstrated the power of the use of telecommunications, social networks, and ad hoc collaboration. Shortly after the Iran demonstrations, both Facebook and Twitter were targeted at the system level with denial-of-service attacks in an apparent effort to silence one Georgian dissident. Conventional information targets in the United States and South Korea were also under attack in the same time frame from unknown sources. The asymmetry was so one-sided that the adversaries were not even visible. Formal government efforts, patriotic individuals aligned with those

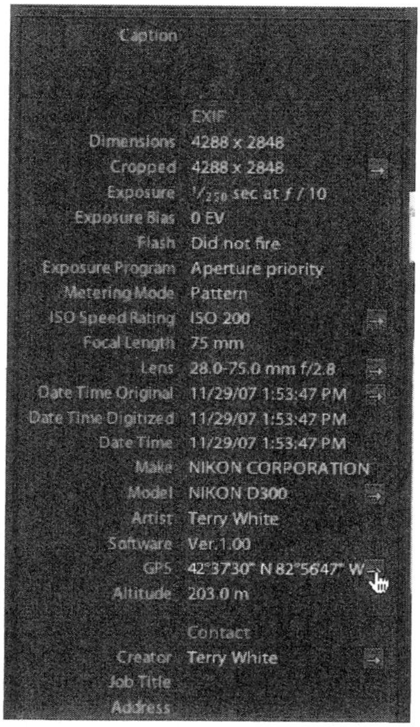

Figure 12.1 Metadata embedded in digital photos can include date, time, GPS location, and altitude.

governments, or a combination of the two were all suggested as sources [4]. Even the number of compromised computers in the July 2009 "bot net" was unclear: estimates ranged from 20,000 to 166,000. Indeed, one main thrust of cyberwar preparedness was to increase the ability of U.S. and allied governments to detect the source and scale of attacks [5].

One highly asymmetric facet of information warfare in the early twenty-first century is the drop in cost for a given capability. Activities that previously required sophisticated manpower and expensive hardware can now be undertaken from a private residence, or even in motion in a passenger vehicle utilizing public mobile data networks. A more general tendency that affects the shape of information warfare is the decentralization of infrastructure. Photographs in the early 1990s were processed and printed in multimillion-dollar facilities. Music was recorded in specialized studios at rates measured in the hundreds or thousands of dollars per hour. Video production and distribution was extremely expensive. Research into specialized databases, such as Lexis/Nexis, occurred at law firms and selected libraries at research universities. In these and other settings, the personal computer and smartphone have made an individual's desktop or pocket equally capable at these formerly expensive activities. The use of well-produced online video to motivate terror group members is one application of this decentralization to asymmetric information warfare.

When capabilities are so easily procured, it is no longer possible to track these "information warfare weapons" the way physical arms and assets are tracked. Given that virtually anybody is capable of creating a propaganda video, or a botnet, or detailed maps using commercial satellite and/or terrestrial images, the relevant question is no longer who *can*, but who *would want to* undertake a given action. Here we move directly into the human landscape where physical sensors and conventional fusion practices are both of limited utility.

12.2.3 Military and Governmental Expectations and Demands

The requirements pull for research in human-centered fusion is evident by multiple programs, conferences, and workshops in this area. Special workshops on soft and hard data fusion were held at the 11th International Conference on Information Fusion (held in Cologne, Germany, on June 30–July 3, 2008) and at the 12th International Conference on Information Fusion held in Seattle, Washington, on July 6–9, 2009 (http://www.isif.org/). These sessions sought to highlight research conducted by government agencies and contractors related to observing and characterizing the human landscape/terrain, utilization of human observers, and fusion of hard and soft data. In addition, a workshop, 2008 Critical Issues in Information Fusion Workshop (http://www.infofusion.buffalo.edu), hosted by the Center for Multisource Information Fusion (CMIF), University at Buffalo/CUBRC, focused on defining basic concepts in hard and soft data fusion and developing foundational process flows and architectures. Details can be found at the referenced Web site.

In addition to theoretical developments, much effort has focused on developing "human terrain mapping" concepts for tactical applications [6, 7]. Marr [6] describes an application of a human terrain study conducted in mid-2007 in Iraq by a unit named Task Force (TF) Dragon (part of the 3rd Heavy Brigade Combat Team). When the task force arrived in Iraq, they discovered that the available data about their assigned area of operation was out of date. The unit proceeded to collect information on their own to understand the human landscape as well as the physical landscape. Similarly the U.S. Army implemented a program called Human Terrain Systems (HTS) [8, 9]. The objectives of HTS are to provide commanders with relevant sociocultural information and knowledge for more effective operations and to harmonize selected courses of action with target area cultural context [9]. At this time, HTS consists of 27 teams deployed in Iraq and Afghanistan. Each HTS team is comprised of five to nine people, including experts on cultural analysis, regional studies, and human terrain analysis.

The Mapping Human Terrain Joint Capabilities Technology Demonstration (MAP-HT JCTD) project (http://tiny.cc/AR7EX) was initiated in 2007 to enable a common operating picture (COP) of the sociocultural terrain and to provide a computer environment to support and improve analysis. The MAP-HT JCTD seeks to develop a toolkit to "collect, consolidate, visualize and understand open-source socio-cultural information in operating areas."

It should be noted that these programs are not without controversy. On October 31, 2007, the American Anthropological Association condemned HTS due to concerns about compromise of ethics, endangerment of research subjects, and potential negative effects on anthropology as an academic discipline [10]. Other academics have labeled HTS concepts as "mercenary anthropology" [11]. Finally, it is clear that military collection of human terrain data is not without danger. On January 7, 2009, Paula Loyd, a Wellesley-educated anthropologist, died in Texas due to injuries sustained by an attacker in Afghanistan. She had been interviewing a local villager who lit a jug of cooking oil and engulfed her in flames. Her death has reverberated throughout academic circles and has heightened fears about the use or misuse of anthropological expertise.

Despite these fears and condemnations, increasing emphasis is being placed by military organizations on understanding the human terrain and collecting and analyzing associated data. New software products are being developed for data analysis, including MapHT by Overwatch Tactical Operations, a division of Textron Systems (http://tactical.overwatch.com/products/mapht.html.) and *Analyst's Notebook* by I2 Inc. (http://www.i2inc.com/products/analysts_notebook/).

12.2.4 Demographics

Whether they are called "oyayubizoku" ("clan of the thumb") in Japan, "digital natives" [12], or the "Net Generation" [13], the generation currently under 30 is argued to access, process, and understand information in markedly different ways from their predecessors. This group played a decisive role in the 2008 U.S. elections, for example, utilizing blogs, online video, social networks, and text messaging to communicate and coordinate. Not long afterward, the same tools played a central role in India's elections. In large measure because of its relationship to technology, that generation is also changing the workplace, personal relationships, dating, and recreation (with online gaming, fantasy sports leagues, and poker, in particular).

Several dynamics of the "digital natives" are relevant to the emerging need for human-centered fusion. First, the ease of online collaboration and interaction has lowered coordination costs: mobilizing a group of three or 3,000 has never been easier, and the underlying assumption is that because people are reachable, they can be counted on to contribute a little or a lot to a project they care about.

Mobilizing people, whether for a party or a cause, is primarily a matter of wanting to. In recent months, for example, the famous professional bicyclist, Lance Armstrong, created instant throngs of 5,000 or more cyclists in a city by simply sending a Twitter message ("tweet") that he is in town and going for a ride. Text-based interactions are so comfortable that the line between friend, acquaintance, and stranger becomes easily blurred. Digital privacy and trust are very much in flux.

Constraints are viewed as artificial. If I want a movie or a song or a piece of information, the reasoning goes, it is a matter of finding it rather than paying for it, and once it is found, I want it right now. Identity is fluid, requiring tools to broadcast these changes to one's network. Because so much of life is lived online and in networks, the customization of computers, environments, and tools is frequent. Identity can become more a matter of what is projected than what one is.

Speed matters; having to wait on a modem-grade connection causes nearly physical pain. Multitasking is common, to the point where the need to prohibit texting and driving is motivating legislation in many states. The speed of change is very real to the digital natives. Within a single family, for example, the ages at which children get mobile phones or Facebook accounts will invariably drop as younger siblings come of age.

12.3 TECHNOLOGY PUSH: ENABLERS FOR HCF

The persistent improvements in line with Moore's law, along with parallel advances in bandwidth, storage, and both human-data and human-computer interaction mean that innovations in both hardware and software are propelling the pace of change in human-centered fusion.

12.3.1 Interconnectivity

Interconnectivity can be measured in both objective and subjective ways. Objectively, the number of Internet connections is increasing, as are the speeds of those connections, even when they are mobile.

- According to the International Telecommunication Union (ITU), the compound annual growth rate for mobile phone subscriptions, worldwide, from 2003 to 2008 was 23%, implying a doubling every 3 years [14].
- Cisco predicts global Internet traffic will grow, driven largely by video, at an annual rate of 40% from 2008 to 2013 [15].
- Facebook grew about 67% in roughly six months in 2009, off a large base, from 150 million to 250 million users [16].

- Japan's KDDI Corporation began delivering consumer gigabit Internet connectivity in late 2008.

Interconnectivity can also be increased subjectively, in part through the spread of standards and other shared understandings regarding technology. Starting in 2002 or so, vendors invested heavily in Services Oriented Architectures, or SOA. This effort involved systematic approaches to standards-based interfaces across operating environments, applications, and data in both technology organizations and business processes. The ideal of SOA is simultaneously to deliver the cost and efficiency benefits of standardization and to support the need for flexibility required by local change and rapid adaptation. Interoperable, reusable services can be made available to support agile development of new capabilities. At the same time, the requirements of management, including security, performance, and cost of ownership, are also supported.

12.3.2 Computational Horsepower

The conventional measure of computational capability remains Moore's law (see Figure 12.2). In desktop and server machines, computing power continues to double roughly every two years, though in mobile applications, power and heat considerations impose different tradeoffs.

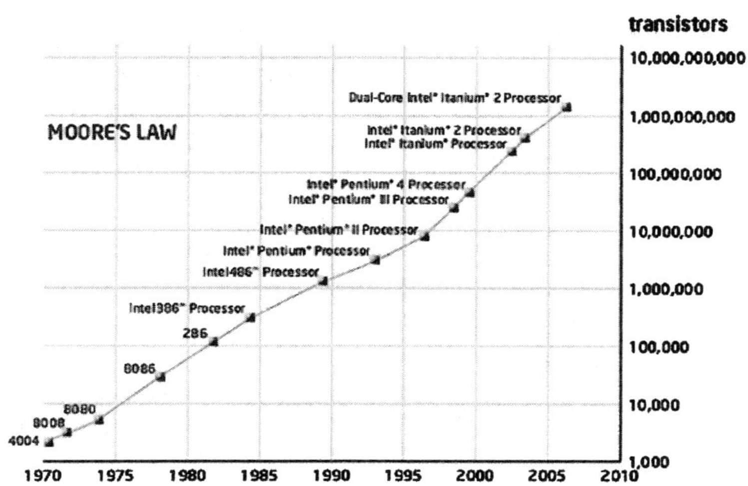

Figure 12.2 Moore's law, originally projected 10 years forward from 1965, has held true for over 40 years. (Source: Intel)

Another way to appreciate the relentless pace of hardware improvement is to look at a representative smartphone, the Nokia N95. It boasts the following specifications:

- 240 x 320 16m color display;
- USB and Bluetooth connectivity;
- 64 MB RAM, 8 GB storage;
- GPS;
- 5 megapixel camera;
- MP3 player, stereo speakers;
- 30 fps VGA video;
- Wi-Fi;
- 3-D graphics accelerator;
- 6.5 hours of talk time;
- 120 grams weight, 90 cc volume.

The phone was announced in 2006 and launched a year later at a U.S. price of $750. For that sum, the buyer basically got the equivalent of an Apple desktop machine (in processing capability, memory, and storage) from only 8 years prior, plus a camera more powerful than anything on the market in 1999, plus a GPS, plus Wi-Fi access—in a package the size of a cigarette pack.

In addition to mobile handsets and PCs, a third platform has emerged as Google, IBM, Amazon, and other vendors are making compute cycles available as a commodity, accessible to anyone on the Internet. In so-called cloud computing, even such computer-intensive applications as video transcoding or protein folding visualizations can be performed using these commodity cycles. Clouds have several desirable characteristics, most notably the ability to scale both up and down as demand merits: idle infrastructure that was bought to accommodate peak loads in an organization becomes a thing of the past, which delivers both cost savings and environmental benefits [17].

12.3.3 Data Structures

As in so many areas, advances in structural data quality are emerging from the top down and the bottom up. From the bottom, the emergence of tagging as a process for metadata generation has moved rapidly, giving everything from press releases to photos crowdsourced categorization. In addition, tools at Amazon (Mechanical Turk), Google (Image Labeler), and elsewhere are capturing human knowledge about images, codifying terms into more structured knowledge than the tag clouds available at Flickr. User rankings, search histories, Bayesian and other filters (such as spam filters), and recommender systems also build on user behavior to increase the quality of future queries.

From the more formal perspective, semantics are being integrated into data warehousing, improving the quality of the metadata, and thus query performance and accuracy. New commercial tools allow organizations to rationalize physical metadata constructs on the basis of commonly used business terminology and to write reusable, target-specific data transformation and business rules [18].

12.3.4 Human-Data Interaction

As we have seen in Chapter 6, the combination of improved graphical processing, improved data-handling practices, and lessons from computer gaming and elsewhere is facilitating a new era in information visualization. Whether in the work of labs, information artisans, or decision-makers striving to cope with information overload, visual tools and methods continue to improve. This process is made possible by improved tools (Google Earth, for one prominent example), inspirational leadership from the likes of Hans Rosling, Edward Tufte, and Jefferson Han, and user demand that information transcend the spreadsheet.

Beyond visualization, the senses of touch and sound are engaged, again following the lead of the gaming world. Haptics, sonification, and other sensory and emotional triggers are finding their way into automobile controls, and other information environments will follow. Chapter 7 describes evolving concepts in sonification. The rapid commercial introduction of devices such as data gloves, head trackers, and wiimotes (for interfacing with wireless game interfaces) will make touch interfaces commonplace. To date, little systematic work has been conducted on exploring how multisensory interfaces will affect interaction with large data sets. Can haptic interfaces allow representation of second-order uncertainty (for example, uncertainty in the error of a quantity) by using a "squishiness" interface (in which an accurate second-order uncertainty is represented by a relatively hard error ellipsoidal surface while a relatively inaccurate second-order uncertainty would be represented by a soft surface)? Can musical harmony be used to define "normal" situations with disharmony, indicating anomalous conditions? Are there possible utilizations of deliberately induced synesthesia to cross sensory representations? This would appear to be a rich area for exploration.

12.3.5 Computer-Enhanced Humanity

The days of a computer being accessible only via a screen, keyboard, or mouse are nearing an end. Combining one's virtual state (restaurant reviews, for example, or traffic updates) with one's physical presence is no longer the work of expensive equipment in research labs. A recent iPhone application overlays augmented reality, turning science fiction into reality in the pursuit of the correct subway entrance. The Nintendo Wii Sports game with its haptic interface has sold nearly 50 million copies.

RFID chips are routinely implanted in animals, and some humans are already experimenting with these implants for identification, access, and other purposes. Cochlear implants electronically improve hearing via direct connection with the auditory nerve, robotic arms are being successfully used as prostheses, and sensing and computing are embedded in the clothing of athletes and firefighters. The possibilities for digital augmentation of human experience, whether through goggles, brain waves (a thought-to-speech translator is still in the laboratory phases), or nerve endings constitute one of the most promising research areas of this century.

Figure 12.3 Concept of 3-D augmented reality.

Figure 12.3 illustrates the concept of augmented reality. The figure shows an experimental environment at The Pennsylvania State University's College of Information Sciences and Technology (IST) Extreme Events Laboratory [19]. A central data analysis facility is located in the IST building on the University Park campus. The analysis facility includes a 3-D visualization facility, multiple large screen displays, 3-D sound capability, and haptic interfaces. Observers about the University Park campus can provide input data to the central analysis facility (acting as soft sensors). In turn, analysis performed at the facility can be transmitted back to an observer (shown wearing the 3-D glasses) allowing models to be projected onto the real world. In this example, information about a hypothesized chlorine plume (due to a chlorine leak in the natatorium) can be sent to the local observer to allow him or her to avoid extensive contamination.

12.4 THE BUSINESS CASE FOR HUMAN-CENTERED FUSION

Human-centered fusion draws on many innovations in business practice from the past 20 years. These include the rise of alternative organizational forms such as Linux and Wikipedia, the rise of free and good-enough alternatives to for-fee products and services, and a rich ecosystem of developers of mobile applications for handheld devices.

12.4.1 Traditional Information Economics

In contrast to physical goods, information goods such as software, movies, and newspapers exhibit several unique and differentiating characteristics [20]. First, because digital copies in particular are nearly free to produce and distribute, many information goods have extremely high fixed (first-copy) costs but low (approaching zero) marginal costs. In this view, the first copy of Microsoft Vista essentially cost billions of dollars; the second copy was the cost of a burned DVD or download.

Second, because the marginal cost of information goods is so low, pricing is determined more by value than by cost plus profit. Because information users have so many different ways of valuing information goods, these goods tend to have many price points in an attempt to appeal to the largest possible number of customers. Student and senior-citizen discounts are common at movies and elsewhere. Gold, silver, and platinum levels of functionality or technical support might be available. Time is a common way to vary pricing: matinees are cheaper movie showings than Friday at 9 p.m., and real-time stock tickers cost more than those with various delays.

Third, information goods are frequently networked in both human and electronic senses. This trait leads to economic externalities: outcomes not accounted for in the price. Negative externalities include the familiar example of pollution. Network externalities can include lock-in: because everyone uses a particular software standard, I have to as well. On the other hand, network externalities can be positive: if I buy the world's first fax machine, it is worthless; but after they become popular, every fax machine sold anywhere in the world makes me a tiny bit better off because now one more person is accessible to my machine.

12.4.2 Emerging Information Economics

Because the Internet lowers coordination costs, large-scale ad hoc collaborations are emerging. Wikipedia and the Linux operating system provide two examples where money does not change hands, yet armies (not mobs) of volunteers have created category-leading products. Facebook confronts a related challenge, expending considerable sums for bandwidth and storage—users upload more than a billion photos to the site every month, for example—and creating unquestionable value for its users, but the company's revenues have yet to keep pace with expenses. Once the link between cost and value is broken, these and other commercial efforts are struggling to find new profit models. Music labels and newspapers also must reconcile value delivery and revenue generation. Cable television and other broadband providers, in contrast, are successfully charging for their services, but face a future in which content (professional football games, for example) might be seen as valuable, while carriage ("dumb pipes") holds less value.

12.4.3 The Emerging Paradox: Competing with "Free"

As the Linux/Wikipedia example illustrates, it can be extremely expensive to compete with free. One estimate values the Linux volunteer contribution at over $10 billion [21]. The cost in bandwidth to duplicate a peer-to-peer file-sharing network would be similarly prohibitive. By not charging, these loose federations give themselves extreme flexibility: user expectations are low, support costs are borne by users who generally operate in self-service mode, and new releases do not need to match investor or customer expectations. Most importantly, the operative currency becomes reputational rather than monetary, at once lowering costs but increasing the difficulty of duplication. For reputation to have value, the conferring institution, whether eBay or a game environment, must itself establish credibility among knowledgeable participants.

As the British sociologist Richard Titmuss showed over 40 years ago, there are markets that function *less* effectively once money is involved; he used human blood donation as a powerful case study [22]. Even now, the quality of Red Cross blood versus purchased plasma is clear. Titmuss' insights have implications for human-centered fusion. As the Internet lowers coordination costs, networks of voluntary contributions become increasingly more feasible, and it is easy to find one trusting complete and anonymous strangers more than erstwhile authorities, in matters of technical support or hotel reviews, for example. As the ability to create voluntary networks increases and the value of the information on those networks improves, established incumbents find themselves struggling to "compete with free."

The asymmetry of information warfare exhibits a similar dynamic. Buying protection from a terror attack is orders of magnitude more expensive than motivating or mounting the attack. The information needs of those preventing things from happening vary considerably from those initiating action. In both the commercial and government settings, to quote John Arquilla and David Ronfeldt, "It takes a network to fight a network" [23]. A key question becomes that of motivation and coordination.

12.4.4 *The Innovator's Dilemma* for Information

Over a decade since its publication in 1997, Harvard professor Clayton Christensen's book *The Innovator's Dilemma* [24] continues to instruct readers on the patterns of innovation and incumbency. In brief, Christensen advances the idea that good management by an incumbent—listening to customers, moving up the value chain to deliver higher profit margins, and continually improving the capability of products and services—might actually cause firms to fail. The fate of such companies of Sears Roebuck, Digital Equipment Corporation, and Link Belt (in cable-actuated excavators) testifies to the broad reach of his insight. In contrast, what he labels "disruptive innovation" enters a market with tiny market share and superficially *less* capability than the incumbents. But disruptive innovations shift the entire premise of the market and reset the cost/value equation. While incumbents often overshoot the needs of the mainstream market with excessive functionality and the attendant complexity, disruptive innovators introduce an alternative value proposition, thereby recasting the business model [24].

One relevant example for human-centered fusion can be found in the history of the U.S. steel industry. In the 1950s, American steel firms capitalized on the buildout necessitated by World War II to dominate world markets. But as Japanese and German mills rebuilt using superior technology, the U.S. firms lost ground. In the 1970s, U.S. firms were confronted with another threat, that of minimills. In contrast to enormous integrated mills that took iron ore, coal, and sand from raw material to finished metal, the minimills remelted scrap steel and thus could operate at a smaller scale. Integrated mills took 9 hours of labor input to produce a ton of steel in 1980, but improved to 3 hours by 1991. Normally such improvement in efficiency would provide a decisive advantage, but the minimills, operating under a different business model, required only 36 minutes of labor per ton in

1995. Integrated mills cost in the vicinity of $6 billion to build, while a minimill could be up and running for about $400 million [24].

In the market, minimills began by offering rebar, which is cheap and undifferentiated and delivers slim profit margins (see Figure 12.4). The incumbents responded not by attacking the insurgent, but ceded the rebar market, concentrating on higher-margin products: rod and bar stock, structural beams, and, at the top end, sheet steel for auto bodies, appliances, and the like. However, over time the minimills learned how to make first rod and bar then structural products, taking advantage of their superior cost model. Up until the end, however, the incumbents appeared to be healthy as the sheet steel market, their last refuge, delivered handsome margins. Bethlehem Steel's market capitalization rose roughly twelvefold from 1986 to 1989, in part because of an investment of over $1 billion in R&D and new equipment, but once Nucor, Chaparral, and the other minimills made the final step to thin-slab casting in the late 1980s, the integrated mills' days were numbered. Because the minimills' advantage was structural, in their businesses' DNA, as it were, the integrated mills eventually had to concede defeat. In less than a generation, the Pennsylvania economy and physical landscape were transformed as Bethlehem, U.S. Steel, Jones and Laughlin, and other firms all either retrenched or closed down completely.

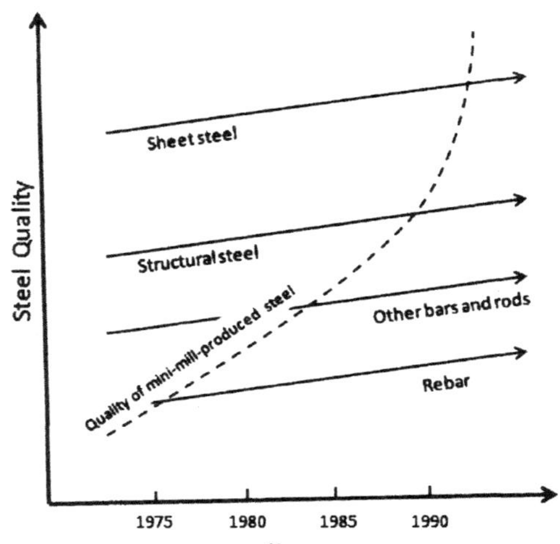

Figure 12.4 Minimills began as a disruptive innovation at the low end of the steel market, but in less than 30 years their inherent cost advantage led them to defeat integrated mills in all product segments.

The current information environment can be seen in parallel to steel. Various government agencies and private enterprises deliver expensive information services at a high level of quality. At the bottom, however, free or low-cost services can be obtained that, while not perfect, might be good enough. The rise of Google at the expense of Lexis/Nexis might be seen as an example. Similarly, satellite images of the earth can be obtained from many sources, from Google to Microsoft (whose Virtual Earth included a photo of an uncloaked propeller of a U.S. *Ohio*-class submarine in dry dock) to SPOT Image, which will deliver satellite photos to order for roughly $10,000. In the realm of mapping, countless low-cost providers exist, most notably OpenStreetMap, which competes directly with the UK Ordnance Survey, which cannot contribute to the project even though it is taxpayer-supported. In images, one could commission a photo of a given artifact, or obtain one free at Flickr, complete with tags (labels) and possibly geotags. Similar to the minimills, the free efforts are getting better and better, leaving incumbent information services with less and less of the market uncontested. Many information providers, from Bloomberg to music labels to newspapers, are confronting similar shifts. The music industry trade association tried suing its customers, which failed to increase purchases of physical CDs, while newspapers are trying to start charging for content after voluntarily giving it away for over a decade.

12.4.5 Three Potential Business Models for Human-Centered Fusion

Recall that human-centered fusion has four components: (1) observing the human landscape, (2) humans as soft sensors and sensor platforms, (3) hybrid computing, and (4) crowdsourcing of analysis. Three potential business models bear on various facets of these components.

Business Model 1: Conventional Organizations

While wikis and bloggers can achieve a lot, they rely on newspapers with expensive foreign bureaus and known brands. Maintaining the model builds on past practice but will require infusions of cash to maintain service levels. Even then, these organizations will face user defections to more nimble competitors. Institutional mass implies slower reaction to market shifts: How fast could *Encyclopaedia Britannica* have added an entry on Sarah Palin after her vice-presidential nomination? Wikipedia had an entry up since October 19, 2005, nearly 3 years before the nomination, and incrementally improved and revised it.

Conventional organizations impart a different kind of credibility to the information they process. Whereas the crowdsourcing model publishes in relatively raw form in order that the crowd can edit and improve the material, established entities such as the *New York Times* filter extensively and then publish. For institutional information, such as GDP figures or stock quotes, authoritativeness is paramount. Thus, the first question in assessing the business model becomes one of assessing what *kinds* of information are in play and whether a crowdsourcing model can even apply [25].

Because, as we have seen, information goods are priced not in relation to cost of production but by value to the user, conventional organizations in various information businesses spend considerable time guarding that value relationship with their customers. They might do this by some combination of proving and improving that value to their users, and creating barriers to entry by contract, technical innovation, network effects (e.g., standards), or lock-in mechanisms such as proprietary interfaces. While proprietary interfaces can be useful, at some point many expensive and rigid platforms become vulnerable to good-enough efforts that utilize an alternative business model: digital photography provides one example.

Business Model 2: E-Lancers

Although the appeal of "the brand called you" and other glorifications of temporary employment has faded somewhat, many people continue to assemble project-based careers rather than commit to individual employers. Buyers of information can capitalize on these networks, in contract photography, for example. What might be some models for human-centered fusion?

Our Man in I-Space

Building on the familiar model of stringers that report only in event of news and cost far less than a dedicated foreign bureau, this model would utilize individuals who would monitor Twitter feeds, blog posts, podcasts, and other relevant Internet sources relevant to some topic. Geography would be largely irrelevant, though domain expertise, potentially including local presence, would be essential. Much as Microsoft built a large and strategically important network of software developers by providing certification, the entity that could provide a similar *Good Housekeeping* seal on information sources would help establish a market among potential buyers for the stringers' services.

The Intell App

To lower the barrier to contribution, an iPhone application could recruit eyes, ears, or brains as necessary, using the GPS information from the phones to task their owners based on a profile. Sample questions might include "What languages do you speak and read?" "What news sources do you track?" and "Have you lived in the area you plan to analyze?" Because the barrier to contribution is low, input would be discounted accordingly but would also have the benefit of being fresh and geographically advantageous. Potentially a game format like America's Army, combined with open-source problems, could serve multiple purposes: recruitment, analysis, and scenario exploration.

Business Model 3: Utilize Existing Volunteers

Whether it is plane-spotting, bird-watching, or wardriving (mapping Wi-Fi networks in a car equipped with a sniffer), many hobbyists devote considerable time and effort to systematic information collection and sharing. The work of such volunteers can be useful in both context-setting (what is the baseline?) and monitoring capacities. Unlike the "Intell app," in which volunteers are solicited, the use of existing information requires seeking out niche and sometimes fringe groups, and playing by their rules.

Structure can be brought to this data in a variety of ways. Even though the promise of information markets for defense analysis suffered a setback when the Policy Analysis Market was shut down in 2003, the fact is that aggregating opinions and predictions can increase accuracy and, if nothing else, stretches the thinking of insiders. Done correctly at aggregate levels ("What will Somalia's GNP do in the next three years?"), prediction markets can motivate a wide variety of independent individuals to contribute informed perspectives. The motivation—reputation among one's peers—already animates both other prediction markets such as Hubdub and leads to such crowd-source successes as Linux.

Alternatively, Infochimps is an open-source data repository for quantitative data sets, with over 5,000 additions as of the time of this writing. A recent sampling included the following:

- Multiple collections of words, whether for crossword puzzles, hyphenation points, or dictionary definitions;
- Stock market data, including daily opening, closing, high, and low prices as well as trading volume for the New York Stock Exchange, the NASDAQ, and other exchanges dating to 1970;
- Game logs and box scores for major league baseball games dating back to 1871;
- Ace Hardware locations.

Metaweb's Freebase project goes a step further, aggregating existing data sets, disambiguating them, and making them searchable utilizing a semantic query language.

In each of these instances, information quality may be a concern, but as with most active open-source efforts, the community will filter out errors in a manner reflecting the community's values. Compared to the high or prohibitive cost of creating such data in traditional channels, learning how and where to use good-enough public sources may become a necessity for many traditional organizations.

12.5 PROSPECTS FOR IMPLEMENTATION

What are the prospects for implementing a human-centered fusion approach in the near future? On one hand, the information flow "plumbing" is near at hand—the rapid evolution of service oriented architectures, Web-based services and programming toolkits such as Adobe Flex, and information-sharing services such as Twitter and Jodange make the infrastructure readily available in the near future. In addition, movements to cloud computing will provide computational "horsepower" to allow nearly anyone to develop and execute complex mathematical models for prediction of physical phenomena such as weather, chemical plume dispersion, and environmental changes. On the other hand, the actual formulation and implementation of appropriate processing flows for human-centered hard and soft fusion will require additional research.

Llinas and Hall [26] have developed two frameworks for understanding the processing flow and algorithms for fusion of hard and soft data summarized here.

1. *Implementation framework:* The first view of the framework provides a basis for algorithm selection, implementation, and demonstration. The framework is shown in Figure 12.5. Three parallel processing flows are considered:
    a. The first flow, shown on the top part of Figure 12.5, addresses the processing of human reports with functions such as message formulation, word-sense disambiguation, automated filtering, soft source characterization, and focus of attention/knowledge elicitation aids.
    b. The second process flow, shown in the middle of Figure 12.5, addresses the typical processes required for ingesting hard sensor data including common functions such as signal conditioning, feature extraction, common referencing, and intersource association.
    c. Finally, the third processing flow, shown at the bottom of Figure 12.5, addresses the types of processes required to address Web-based

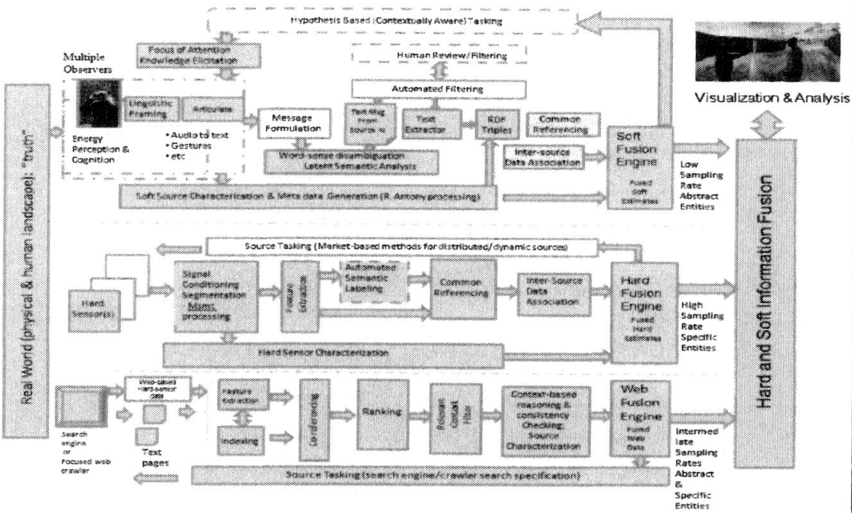

Figure 12.5 A framework for automation.

information, using search engines in effect as "Web-source observers." This framework shows the top level functions for processing hard and soft data (and information from the Web). At any point in the processing flow, one or more humans could participate. For example, filtering of textual data could be performed by automated text-processing software, by human inspection, or a combination of the two. This framework could assist as a basis for implementation of a software infrastructure to support algorithm implementation, test, and evaluation.

2. *Functional hierarchy:* Another way to develop a functional framework (and identify relevant models and algorithms) is to follow the Joint Directors of Laboratories (JDL) Data Fusion Process Model [27–28]. This model implies an analog between traditional fusion processing (at the JDL level 0 and level 1 subprocesses) for physical sensors observing physical targets and creating similar subprocesses and functions for soft sensors observing the human landscape. The result of this approach is shown in Figure 12.6.

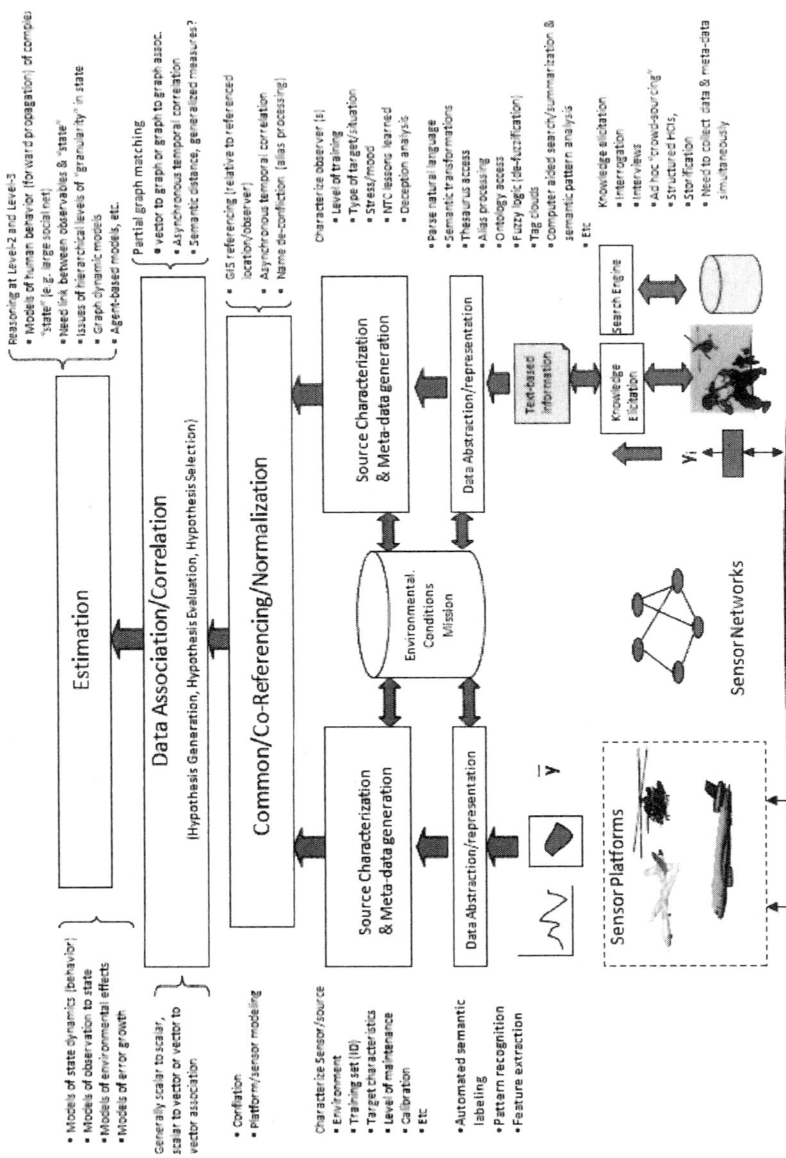

Figure 12.6 Information fusion hierarchy (JDL Levels 0 and 1) for hard and soft data.

In Figure 12.6, the left side of the figure shows the level 0 and level 1 processes for translating traditional hard sensor data (e.g., signal and image data) about physical targets to state vectors that represent or characterize the targets (e.g., position, velocity, identity, and attributes). The types of functions include data abstraction and representation, source characterization and metadata generation, coreferencing, data association and correlation, and finally state estimation. Types of algorithms and models are shown on the far left side of the figure.

By contrast, the right hand side of the figure shows the parallel functions and algorithms for soft sensors observing the human landscape. In this case much of the data involves text-based information, which still requires functions such as data abstraction, source characterization, coreferencing, data association, and correlation and estimation. However, the specific types of algorithms shown on the far right side of the figure are different than those used for processing hard sensor data. This is to be expected. It should also be clear that, as in processing of hard sensor data for the physical landscape, there are no "magic algorithms" that address all aspects of modeling and prediction for the human landscape.

While these frameworks demonstrate some of the processing flows and functions required to automate the human-centered fusion process, much work remains to define, implement, and evaluate specific algorithms and techniques.

12.6 CONCLUSION

Human-centered fusion is pushed by rapid innovations in technology (microsensors, ubiquitous cell phones, increasing connectivity and bandwidth, and new advanced human-computer interfaces) and pulled by a new generation of digital natives who interact with the Web and each other in dynamic ad hoc distributed ways and pulled by the need to understand the human landscape as well as the physical landscape. Together, these forces will affect the economics of the information supply chain, from creation to distribution and use. Waltz [29] correctly noted that warfare has evolved from a focus on the number of combatants to capabilities of weapon systems to logistics to information collection, dissemination, and understanding. Whether in business or military applications, such trends will continue. As global data collection and dissemination increasingly face "free" and readily available alternatives, decisive advantages must be sought in how to effectively utilize limited human resources: as observers, human computers and analysts, and collaborators.

It is interesting to note that some data fusion researchers appear to be experiencing a career trajectory similar to that observed in some well-known experts in artificial intelligence and computational science, such as Clifford Stoll [30]. At the beginning of their career, they are impressed by the capabilities of computers, saying in effect that computers are wonderful and can be ultimately made to perform tasks commonly performed by humans. Later in mid-career, they express frustration at the large challenges involved and complain that if only they had sufficient computer memory and computational speed, the problem could be solved. Much later in their careers, they note the impressive ability of humans, saying, in effect, aren't humans wonderful?

We suggest that human-centered information fusion will contain multiple challenges for many years to come. On the demand side, people with access to new technologies and ideas will grow more capable, potentially becoming even more difficult to understand and model. Coordination costs are dropping, making it easier for groups to organize on an often informal basis. Finally, as we have seen, many weapons available to these groups are difficult to track, making the threat of different forms of insurgenceny and disruption nearly ubiquitous.

Inside the information fusion process, meanwhile, human attitudes, habits, and capabilities must undergo substantial shifts to accommodate these new needs and threats. Computational capability continues to accumulate, but it still requires human models, algorithms, and interpretation to generate useful action. Complexity scales nonlinearly, making both large organizations and sophisticated models and processes vulnerable to gridlock. In the end, human-centered fusion will likely require new people, new processes, and new organizations to invent and explore its potential.

References

[1] Raento, M., A. Oulasvirta, and N. Eagle, "Smartphones: An Emerging Tool for Social Scientists,""*Sociological Methods Research,* Vol. 37, No. 3, 2009, pp. 426–454.

[2] Francis, D., "P&G's Army of 'Moms,'" *Financial Post*, July 7, 2006.

[3] Rappeport, A., "Whole Foods CEO Apologizes for 'Error,'" *CFO*, July 18, 2007.

[4] Gorman, S., and E. Ramstad, "Cyber Blitz Hits U.S., Korea," *Wall Street Journal,* July 9, 2009.

[5] Bain, B., "Intell Chief: Source of Cyberattacks Still Unknown," *Federal Computer Week*, July 22, 2009.

[6] Marr, J., et al., "Human Terrain Mapping: A Critical First Step to Winning the COIN Fight," *Military Review*, Vol. 88, 2008.

[7] McFate, M., "Anthropology and Counterinsurgency: The Strange Story of Their Curious Relationship," *Military Review*, Vol. 21, 2005.

[8] Gonzales, R., "The Rise of the Human Terrain System," *Anthropology Today*, Vol. 23, No. 6, 2007, pp. 21–22.

[9] Finney, N., *Human Terrain Team Handbook*, 2008.

[10] http://blog.aaanet.org/2007/11/07/aaa-board-statement-on-hts-2/.

[11] Rohde, D., "Army Enlists Anthropologists in War Zones," *New York Times*, October 5, 2007.

[12] Palfrey, J., and U. Gasser, *Born Digital: Understanding the First Generation of Digital Natives,* New York: Basic Books, 2008.

[13] Tapscott, D., *Grown Up Digital: How the Net Generation Is Changing Your World,* New York: McGraw-Hill, 2008.

[14] http://www.itu.int/ITU-D/ict/.

[15] Cisco Visual Networking Index: Forecast and Methodology, 2008–2013, http://www.cisco.com/en/US/solutions/collateral/ns341/ns525/ns537/ns705/ns827/white_paper_c11-481360_ns827_Networking_Solutions_White_Paper.html.

[16] Figures from Facebook corporate Web site, http://www.facebook.com.

[17] "Corporate IT: Highs and Lows," *The Economist*, October 23, 2008.

[18] http://www.expressor-software.com/.

[19] Hall, D. L., and S. Aungst, "The Use of Soft Sensors and I-Space for Improved Combat ID," chapter 10 in *Human Factors Issues in Combat Identification*, D. Andrews, R. Herz, and M. Wolf, (eds.), Burlington, VT: Ashgate, 2009, pp. 161–170.

[20] Varian, H., and C. Shapiro, *Information Rules,* Boston, MA: Harvard Business School Press, 1998.

[21] McPherson, A., B. Proffitt, and R. Hale-Evans, "Estimating the Total Development Cost of a Linux Distribution," The Linux Foundation, October 2008.

[22] Titmuss, R., *The Gift Relationship,* New York: The New Press, 1997 reprint.

[23] Arquilla, J., and D. Ronfeldt, *Swarming and the Future of Conflict,* Santa Monica, CA: RAND Corporation, 2000.

[24] Christensen, C. M., *The Innovator's Dilemma: When New Technologies Cause Great Firms to Fail,* Boston, MA: Harvard Business School Press, 1997.

[25] Shirky, C., *Here Comes Everybody,* New York: Penguin, 2008, pp. 81 ff.

[26] Hall, D. L., and J. Llinas, Interim technical report for Modeling and Mapping of Human Source Data, report prepared for phase II of the Modeling and Mapping of Human Source Data project (contract no. W 911NF-07-1-0168) funded by the Army Research Office, September 2009.

[27] Hall, D. L., and S. McMullen, *Mathematical Techniques in Multisensor Data Fusion*, 2nd ed., Norwood, MA: Artech House, 2004.

[28] Kessler, O., et al., "Functional Description of the Data Fusion Process," technical briefing prepared for the Office of Naval Technology, November 1991.

[29] Waltz, E., *Information Warfare Principles and Operation*, Norwood, MA: Artech House, 1998.

[30] Stoll, C., *High Tech Heretic: Why Computers Don't Belong in the Classroom and Other Reflections by a Computer Contrarian*, New York: Doubleday, 1999.

About the Authors

Dr. David L. Hall is a professor in The Pennsylvania State University College of Information Sciences and Technology (IST), where he leads the Center for Network Centric Cognition and Information Fusion. Previously he was the associate dean for Research and Graduate Programs in Penn State's College of Information Sciences and Technology, where he was responsible for overseeing IST's Ph.D. and master's degree programs as well as research grant administration and leadership of IST faculty research efforts. Dr. Hall joined IST from The Pennsylvania State University's Applied Research Laboratory (ARL) where he served as associate director and senior scientist. At ARL, he oversaw the 150-person Information and Network Systems Office, comprised of four divisions: information science and technology, navigation research and development, systems and operations automation, and communications science and technology. He has more than 25 years of experience in research, research management, and systems development. He is the author of more than 200 papers, reports, books, and book chapters, and he has delivered numerous lectures on his research, research management, and artificial intelligence. His book, *Mathematical Techniques in Multisensor Data Fusion*, has been used as a text at the State University of New York–Buffalo, George Mason University, Colorado State University, the U.S. Air Force Institute of Technology, and the Naval Post Graduate School. Dr. Hall has been named as an IEEE fellow for his contributions to data fusion and he was awarded the DoD Joe Mignona national data fusion award for his contributions and leadership in data fusion.

Dr. John M. Jordan is a clinical professor of Supply Chain and Information Systems at the Smeal College of Business, Penn State University. Dr. Jordan received a Ph.D. in American studies from the University of Michigan, an MAR from Yale University, and a B.A. in political science and history from Duke University. His research focuses on emerging technologies and their impact on business strategy, design, and practice. His Ph.D. dissertation, on the wide use of engineering images and metaphors in American society in early twentieth-century America, was published by University of North Carolina Press. In addition to placing articles in *American National Biography*, *Technology and Culture*, and other professional publications, Dr. Jordan has written a monthly emerging technologies newsletter since 1997 and reviewed books for both *Harvard Review* and *Upside*. Dr. Jordan has also been profiled in the *Wall Street Journal, Fast Company, Investor's Business Daily,* and *The International Herald Tribune.* He teaches information technology strategy along with innovation and entrepreneurship in the MBA and undergraduate programs at The Pennsylvania State University. Before coming to Smeal, he combined business and technology research with client responsibilities in the Office of the Chief Technologist at Capgemini, the Center for Business Innovation at Ernst & Young, and the Applied Technology Group at Computer Sciences Corporation. He has guest-lectured at the business schools of MIT, the University of Michigan, and University of Southern California. Prior to entering consulting, he won teaching awards at Harvard University and the University of Michigan. He has served on the advisory boards of the DEMO emerging technologies conference, the Gartner G2 industry service, and several startup companies.

Index

The Artech House Electronic Warfare Library

Electronic Intelligence: The Analysis of Radar Signals, Second Edition, Richard G. Wiley

Electronic Warfare for the Digitized Battlefield, Michael R. Frater and Michael Ryan

Electronic Warfare in the Information Age, D. Curtis Schleher

Electronic Warfare Target Location Methods, Richard A. Poisel

EW 101: A First Course in Electronic Warfare, David Adamy

EW 103: Tactical Battlefield Communications Electronic Warfare, David Adamy

Foundations of Communications Electronic Warfare, Richard A. Poisel

High-Level Data Fusion, Subrata Das

Human-Centered Information Fusion, David L. Hall, John M. Jordan

Information Warfare and Organizational Decision-Making, Alexander Kott, editor

Information Warfare Principles and Operations, Edward Waltz

Introduction to Communication Electronic Warfare Systems, Richard A. Poisel

Knowledge Management in the Intelligence Enterprise, Edward Waltz

Mathematical Techniques in Multisensor Data Fusion, Second Edition, David L. Hall and Sonya A. H. McMullen

Modern Communications Jamming Principles and Techniques, Richard A. Poisel

Principles of Data Fusion Automation, Richard T. Antony

Stratagem: Deception and Surprise in War, Barton Whaley

Statistical Multisource-Multitarget Information Fusion, Ronald P. S. Mahler

Tactical Communications for the Digitized Battlefield, Michael Ryan and Michael R. Frater

Target Acquisition in Communication Electronic Warfare Systems, Richard A. Poisel

For further information on these and other Artech House titles, including previously considered out-of-print books now available through our In-Print-Forever® (IPF®) program, contact:

Artech House
685 Canton Street
Norwood, MA 02062
Phone: 781-769-9750
Fax: 781-769-6334
e-mail: artech@artechhouse.com

Artech House
16 Sussex Street
London SW1V 4RW UK
Phone: +44 (0)20-7596-8750
Fax: +44 (0)20-7630-0166
e-mail: artech-uk@artechhouse.com

Find us on the World Wide Web at: www.artechhouse.com